Physician's Collection Association

The Medical and Surgical Directory of Cook County, Illinois, 1888-89

Physician's Collection Association

The Medical and Surgical Directory of Cook County, Illinois, 1888-89

ISBN/EAN: 9783744783262

Printed in Europe, USA, Canada, Australia, Japan

Cover: Foto ©berggeist007 / pixelio.de

More available books at **www.hansebooks.com**

This bea[illegible] m the city bearing that name. It is [illegible] on for the comfort of guests. See advertisement on page 131.

McIntosh Battery AND Optical Co.

MANUFACTURERS OF THE CELEBRATED

McINTOSH GALVANIC AND FARADIC

BATTERIES

Table, Office and Family Batteries, Electrodes, Electric Bath Apparatus, Statical Electric Machines, Stereopticons, Solar, Monocular and Binocular Microscopes, and all kinds of Philosophical Electrical Apparatus. We invite special attention to our

Combined Galvanic and Faradic Batteries.

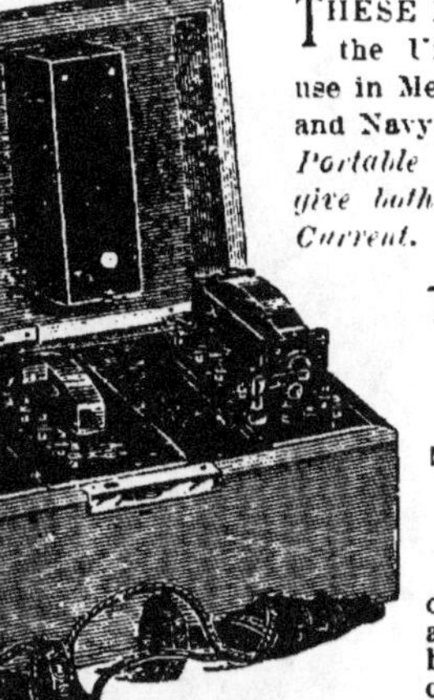

THESE Batteries have been adopted by the United States Government for use in Medical Department of the Army and Navy. They are *the first and only Portable Batteries ever invented which give both the Galvanic and Faradic Current.*

Two Distinct Batteries in One Case.

No Physician can afford to be without one.

Our Batteries are constructed on an improved plan. The zincs and carbons are fastened to hard rubber plates in sections of six each. The cells are composed of one piece of hard rubber, and are made in sections of six each, with a drip-cup. This is the only Battery in which the zinc and carbon plates can be kept clean and always in order by simply rinsing them. An extra large cell is added to the combined batteries for the purpose of producing the Faradic current. All the metal-work is finely nickel-plated and highly polished, and every part is put together so that it can be easily replaced by the operator.

OUR ILLUSTRATED CATALOGUE, a handsome book giving full description of all our goods, and other valuable information, sent FREE on application.

McINTOSH BATTERY & OPTICAL CO.,

141 & 143 WABASH AVENUE. CHICAGO, ILL.

WOMAN'S SCHOOL
OF
OPTICS AND OPHTHALMOLOGY,
CHICAGO.

A POST GRADUATE COURSE FOR WOMEN.

DIDACTIC AND CLINIC INSTRUCTION GIVEN:

In detecting and Measuring Errors of Refraction: In Prescribing the Correcting Lenses for each Eye; In Prescribing, by Measurements, the Frame to fit each Face; In Neutralizing the Lenses; In Verifying the Prescription after it is filled.

For further information address

DR. FANNIE DICKINSON,

Rooms 208-209, 70 State Street, CHICAGO.

Dr. VON GRAEFE, of Berlin, Germany, "*Father of Ophthalmology*," daily devoted the first hour in his office to the examination and treatment of the eyes of the poor.

CHICAGO SPECTACLE CLINIC

Gives, from 9 to 10 A. M., free of Charge, a Professional Examination of the Eye to those persons who cannot afford to pay for examination and glasses. Glasses given free of charge only on written order of the Chicago Charity Organization.

Rooms 208, 209, Bay State Building, 70 State Street, Chicago.

THE WOMAN'S MEDICAL COLLEGE

OF CHICAGO,

335 SO. LINCOLN STREET,

Opposite the County Hospital.

The **Regular Session** begins the **First Tuesday in September,** and continues Thirty-one Weeks. Clinical advantages unsurpassed.

FACULTY.

WM. H. BYFORD, A. M., M. D., PRESIDENT, Professor of Gynecology.

W. GODFREY DYAS, M.D., F.R.C.S., Professor Emeritus of Theory and Practice of Medicine.

G. C. PAOLI, M.D., Professor Emeritus of Materia Medica and Therapeutics.

T. DAVIS FITCH, M.D., Professor Emeritus of Gynecology.

R. G. BOGUE M.D., Professor Emeritus of Surgery.

CHARLES WARRINGTON EARLE, A.M., M.D., TREASURER, Professor of Diseases of Children and Clinical Medicine.

ISAAC N. DANFORTH, A.M., M.D., Professor of Renal Diseases.

HENRY M. LYMAN, A.M., M.D., Professor of Theory and Practice of Medicine.

DANIEL R BROWER, M.D., Professor of Diseases of the Nervous System and Clinical Medicine.

SARAH HACKETT STEVENSON, M.D., Professor of Obstetrics.

DAVID W. GRAHAM, A.M., M.D., Professor of Surgery.

WM. J. MAYNARD, A.M., M D., Professor of Dermatology.

WM. T. MONTGOMERY, M.D., Professor of Ophthalmology and Otology.

E. FLETCHER INGALLS, A.M., M.D., Professor of Diseases of the Chest and Throat.

F. L. WADSWORTH, M.D., Professor of Physiology.

MARIE J. MERGLER, M.D., SECRETARY, Professor of Gynecology.

EUGENE S. TALBOTT, M.D., D.D.S., Professor of Dental Surgery.

JEROME H. SALISBURY, A.B., M.D., Professor of Chemistry and Toxicology.

MARY E. BATES, M.D., Professor of Anatomy.

JOHN A. ROBISON, M.D., Professor of Materia Medica and Therapeutics.

MARY H. THOMPSON, M.D., Clinical Professor of Gynecology at the Hospital for Women and Children.

ELIZA H. ROOT, M.D., Professor of Hygiene and Medical Jurisprudence.

FRANK CARY, M.D., Professor of Pathology and Director of the Pathological Laboratory.

LECTURERS AND ASSISTANTS.

ROBERT S. HALL, M.D., Clinical Lecturer on Midwifery, and in charge of Outside Obstetrical Department.

HOMER M. THOMAS, M.D., Lecturer on Diseases of the Chest and Throat.

ANNETTE S. DOBBIN, M.D.,
ELLEN H. HEISE, M.D.,
RACHEL HICKEY, M.D., } Assistants to the Chair of Anatomy.

MARY MIXER, M.D., Lecturer on Histology and Director of the Pathological Laboratory.

DEREXA N. MOREY, B.S., M.D., Assistant to the Chair of Physiology

M. L. BUSH, College Clerk and Janitor.

For announcements address the Secretary,

MARIE J. MERGLER, M. D.,

29 Waverly Place, Chicago, Ill.

Chicago Medical Journal and Examiner.

ESTABLISHED IN 1844.

S. J. JONES, M. D., LL. D., EDITOR.

A MONTHLY MEDICAL JOURNAL,

PUBLISHED UNDER THE AUSPICES OF THE

CHICAGO MEDICAL PRESS ASSOCIATION.

FIFTY-SEVENTH VOLUME.

SUBSCRIBE FOR IT.

TERMS:

One Year, **Three Dollars, in Advance.**
Six Months, **One Dollar and a Half, in Advance,**
Single Copies, **Twenty-Five Cents.**

It contains **sixty-four pages** of reading matter, exclusive of advertisements, and is issued on the first day of each month.

The matter consists of Original Articles from the pens of some of the best known members of the Profession, Foreign and Domestic Correspondence, Reports of Societies and Associations, Reviews and Editorials, and Items of Interest to Physicians.

Volumes begin with January and July numbers.

Subscriptions received at any time, and back numbers furnished, when practicable, to complete the current volume.

Specimen copies and rates for advertising department furnished on application.

Address

CHICAGO MEDICAL JOURNAL AND EXAMINER,

7 and 9 Jackson Street,

ARGYLE BUILDING. *CHICAGO, ILLINOIS.*

STUTTERING. STAMMERING.

DOCTOR

E. L. RIVENBURGH,

Proprietor of the

WORLD'S VOCAL INSTITUTE.

S. E. cor. Washington and Fifth Av.,

Rooms 16 & 17,

—CHICAGO.—

☞Stuttering, stammering and all speech impediments cured. I remove the *cause* of stammering, and therefore the cure is permanent. The only *scientific* and *successful* method practiced. I *warrant* to cure the worst case of stammering in a few weeks. If you have tried to be cured and are not, remember I *guarantee* to cure. I have made the treatment of Stammering and all Defects in Speech a *specialty* for the past 17 years, and have **cured over 1,600 cases.** My method has never failed. Entrance Nos. 168 & 170 Washington St. and 105 Fifth Ave., Chicago.

IT IS WONDERFUL, YET IT'S TRUE.

From the Chicago Tribune.—Last week (Tuesday) a young man arrived from Portland, N. Y. He stammered at a fearful rate. Indeed, it took him longer to tell his name than it would a quick-tongued man to read a chapter in one of Reade's novels. The same young man was met yesterday, and was heard with a number of other young men and ladies who have been afflicted in a similar manner, and who are attending Dr. Rivenburgh's Vocal institute, to speak out fluently, and read without any perceptible hesitancy in speech. Each and every one under the doctor's treatment is loud in his praise, and advises any one who has any speech impediments to seek Dr. Rivenburgh for relief. His Vocal institute is located at 168 and 170 Washington St.

CHICAGO, ILL., Feb. 1, 1888.—I am willing to testify that Dr. E. L. Rivenburgh, specialist in the treatment of stammering, is and has been entirely successful in several cases that have come to my notice. In one case a pupil of mine, who did not recite orally during his school course, by reason of an infirmity of speech now speaks readily and easily after a short treatment by Dr. Rivenburgh. The doctor claims that he can wholly cure any case, and I believe he can.

A. R. SABIN, Asst. Supt. Schools, Chicago.

FIRST-CLASS IN EVERY RESPECT.

CHIROPODIST AND MANICURE.

OPERATING ROOMS,

109 WABASH AVE.,

(Basement.)

Under the personal supervision of Prof. William Emanuel—a gentleman of many years experience in his profession. Ladies and gentlemen can depend upon receiving the most careful and skillful treatment, and the most gentlemanly consideration.

Hours, 8 A. M. to 8 P. M.

Every Facility Afforded for the Personal Comfort of Patrons.

MAGNETIC AND MASSAGE TREATMENT.

ESTABLISHED 1881.

Treatment of Ladies a Specialty.

HENRIETTA NIX, Masseuse.

407 W. MADISON ST.

PROFESSIONAL REFERENCES.

C. W. Earle, M. D. A. B. Wescott, M. D. Milton Jay, M. D. Adam Miller, M. D. C. A. Dewey, M. D.

PHYSICIANS AND DENTISTS,

IF YOU HAVE ANY COLLECTIONS PLACE THEM IN THE HANDS OF THE

Physicians' Collection Association.

You will receive efficient service at the

LOWEST POSSIBLE RATES.

Send a Postal Card and our Representative will call on you.

Rooms, 111 and 112 Illinois Bank Building,

115 DEARBORN ST., CHICAGO.

See pages 208 and 209 of this Directory.

THE

CHICAGO COLLEGE

OF

OPHTHALMOLOGY AND OTOLOGY,

(INCORPORATED JAN. 25, 1878.)

Located at

527 & 529 State, Cor. Harmon Ct.

The regular term of lectures at the Chicago College of Ophthalmology and Otology will begin the third Saturday in September of each year, continuing six months.

The College is located on the corner of State street and Harmon Court; and is so situated as to make it convenient for students who are attending other medical colleges, as well as to make the clinical advantages unsurpassed.

The lecture room is comfortable, well lighted and ventilated, thus making it a desirable place for students to prosecute their studies. Students will find suitable boarding places within two or three blocks of the college. The building is so arranged as to lend every convenience desirable to the student.

The only Eye and Ear College in the Northwest where a special and complete course of lectures on the eye and ear are given, with all the operations on the living subject. The lectures will be illustrated dy drawings, models, operations, on the human subjeot and animals, also by a weekly clinic of hospital and out patients. Students are also practiced in the use of the ophthalmoscope.

The opportunities thus afforded are abundantly ample for the acquirement of a practical knowledge of this important branch of medical science.

All information regarding the College may be obtained by addressing either

HENRY OLIN, M. D.,
President and Dean of Faculty,
163 STATE ST,

—OR—

J. B. McFATRICH, M. D.,
Secretary,
126 STATE ST.

The Clark & Longley Co.,

PRINTERS,

308-316 DEARBORN STREET,

CHICAGO.

LARGEST ORDERS

Handled with Promptness.

· DR. W. E. DODDS, ·

89 RANDOLPH STREET,

HOURS, 8.30 A. M. TO 6 P. M.

Special attention given to the Treatment of Genito-Urinary Organs, Rectum and Kidneys.

Hemorrhoides and Fissures treated successfully without the use of the knife.

MRS. E. TOWN'S

MASSAGE BATHING PARLORS AND SWEDISH CURE

78 STATE ST., ROOM 40.

New and Commodious Parlors

—FOR—

ELECTRIC AND MAGNETIC TREATMENT AND SEA SALT

BATHS.

FIRST-CLASS IN EVERY RESPECT.
SPECIAL ATTENTION GIVEN TO LADIES AND CHILDREN.

JOSEPHINE BROWNLY, 350 W. MADISON ST.

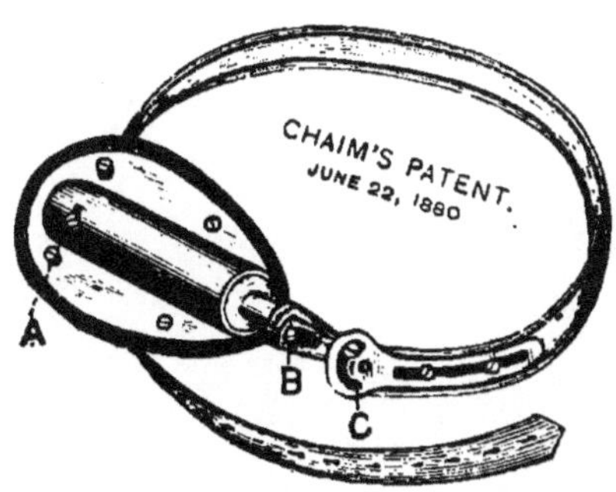

ADOLPH CHAIM, Graduate Bandagist and Instrument Maker. Inventor of the Chaim Truss. Special attention paid to Derformities and Ruptures; also, manufacturer of all kinds of Trusses, Abdominal Supporters, Artificial Legs, Elastic Stockings, Deformity Apparatus, Electric Batteries, Artificial Noses, etc.

70 MADISON ST.

S.W. cor. of State. **CHICAGO.**

Telephone 752.

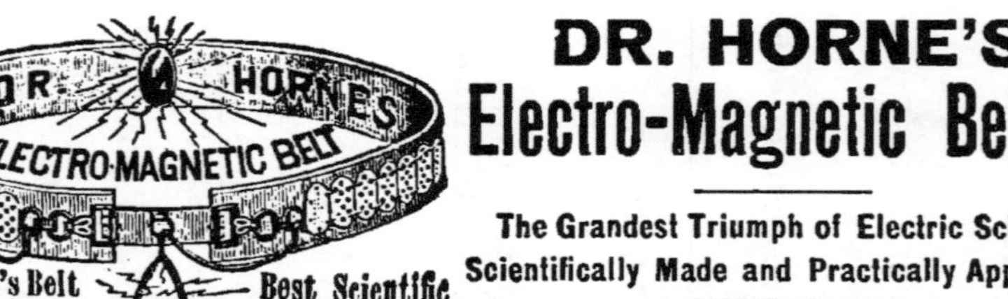

Gentlemen's Belt with Electric Suspensory.

Best Scientific Medical Belts.

DR. HORNE'S
Electro-Magnetic Belts!

The Grandest Triumph of Electric Science—Scientifically Made and Practically Applied.

DISEASES CURED WITHOUT MEDICINES.

IT WILL CURE YOU Have you Pains in the Back, Hips, Head or Limbs, Nervous Debility, Lumbago, General Debility, Rheumatism, Paralysis, Neuralgia, Sciatica, Diseases of Kidneys, Spinal Diseases, Torpid Liver, Gout, Exhaustion, Emissions, Asthma, Heart Disease, Dyspepsia, Constipation, Erysipelas, Indigestion, Weakness, Impotency, Catarrh, Piles, Epilepsy, Dumb Ague, Diabetes, Hydrocele, Blood Diseases, Dropsy, etc., then this belt is just what you need.

Electricity Instantly Felt! Can be applied to any part of the body. Whole family can wear it. It electrifies the blood and cures **WHEN ALL ELSE FAILS.**

TESTIMONIALS Every one genuine and used by permission. **NOTE** the following who have been **CURED:**—A. J. Hoagland, R. S. Parker and J. M. Haslett, all on Board of Trade, Chicago; A. Gregory, commission merchant, Stock Yards; Budd Doble, the great horseman; Col. Connelly, of the *Inter Ocean*; G. W. Bellns, M. D., Mormontown, Iowa; Lemuel Milk, Kankakee, Ill.; Judge I. N. Murray, Naperville, Ill.; E. L. Abbott, supt. city water works, South Bend, Ind.; Robt. R. Sampson, Chicago post office; L. D. McMichael, M. D., Buffalo, N. Y.—"Your belt has accomplished what no other remedy has: steady nerves and comfortable sleep at night." Robt. Hall, alderman, **150** East 39th Street., New York—and thousands of others.

Dr. HORNE'S ELECTRO MAGNETIC BELT is superior to all others—currents of electricity are strong or mild as the wearer may desire; produces a continuous current; conveys electricity through the body on the nerves. It cures diseases by generating a continuous current of electricity (**10** or **12** hours out of **24**) throughout the human system, allaying all nervousness immediately, and producing a new circulation of the life forces—the blood, imparting vigor, strength, energy and health, when all other treatment has failed. The merits of this scientific Belt are being recognized and indorsed by thousands whom it has cured.

REFERENCES:—Any bank, commercial agency or wholesale house in Chicago; wholesale druggists, San Francisco and Chicago. ☞ Send stamp for **112** page Illustrated pamphlet.

DR. W. J. HORNE, Inventor and Manufacturer, **191** Wabash Avenue Chicago.

MRS. EX-PRESIDENT GRANT: "The piano purchased from you five years ago still gives the greatest satisfaction, and having been performed upon by some of the first musicians of Washington, their opinion is qualified approbation for the Bradbury."

MRS. EX-PRESIDENT HAYES: "It is a remarkably fine instrument in quality of tone, finish and touch, and everything that goes to make a truly first-class piano, and gives entire satisfaction in every respect."

DR. T. DEWITT TALMAGE: "It is the pet of our household—should have no faith in the sense or religion of any who does not like Bradbury."

MAJOR GENERAL O. O. HOWARD: "Mrs. Howard and myself cannot speak too highly, or recommend too strongly the beautiful Bradbury piano purchased from you—all that combines to make in every sense a first-class piano is combined in this."

ADMIRAL D. D PORTER: "The Bradbury piano I purchased from you eight year ago has been used almost constantly, and has been tested by the best musicians, and pronounced one of the best they ever saw in finished workmanship and fine quality of tone."

SOLD ON EASY PAYMENTS.

FREEBORN G. SMITH,

MANUFACTURER.

WESTERN BRANCH, 210 STATE STREET.

SPIRAL SPRING CARRIAGES

THE PHYSICIAN'S FRIEND.

INVENTED BY A PHYSICIAN.

ENOCH SCOTT & CO.,

375 WABASH AVENUE,

CHICAGO, ... ILLINOIS.

GEO. C. DARCHE,

MANUFACTURER AND DEALER IN

ELECTRIC GOODS,

MEDICAL INSTRUMENTS,

Batteries, Electric Bells, Burglar Alarms, Electric Gas Lighting, Speaking Tubes and Telephones.

35 AND 37 S. CLARK STREET,

CHICAGO.

Private phones and bells from drug stores to physicians' offices a specialty.

ESTIMATES FURNISHED. TELEPHONE 5687.

ESTABLISHED IN { NEW YORK, 1858. CHICAGO, 1869.

H. J. EDWARDS & SONS,

Manufacturers and Dealers in

ALL THE LATEST STYLES OF

FINE CARRIAGES

—AND—

SLEIGHS,

Edwards' Jump-Seat Buggies, Doctors' Phaetons, Harness, Etc.

292 & 294 WABASH AVE.,

CHICAGO.

JUSTIN HAYES, M. D.,

Removed to 240 Wabash Ave. Take Elevator.

SPECIALTY, TREATMENT OF

Chronic and Nervous Diseases.

OFFICE HOURS.
10 A. M. 2:30 P. M. Chicago.

Experienced Electrical Operators in Attendance from 8 A. M. to 6 P. M.

We use the best medical means known to the profession, including Galvanism, Faradism and Franklinism, with the best medical apparatus known. The Electro-Thermal Bath, as invented by Dr. Justin Hays, receives both the Faradic and Galvanic currents by telegraphic connections. The new Static Electrical Machines are a wonderful power in hastening the recovery of the patient.

Prominent among the diseases treated are ***Diseases of Women;*** Ovarian and Fibroid Tumors successfully treated with electricity during the last twenty-four years. Palsy, Neuralgia, Rheumatism, Spinal diseases, ***Diseases of the Heart***, etc.

Home for care, treatment and board at Western Springs—40 minutes' ride from C., B. & Q. Union Depot, corner Canal and Adams streets.

—THE—

AMERICAN DEPURATING BATH

WILL BE FOUND

Superior to anything ever offered for the cure of all ***CONTAGIOUS DISEASES, ACUTE AND CHRONIC.*** *All diseases of the blood,* ***Rheumatism,*** *Fevers of all kinds, and diseases of the*

NERVOUS SYSTEM.

Recommended by Leading Physicians of the City.

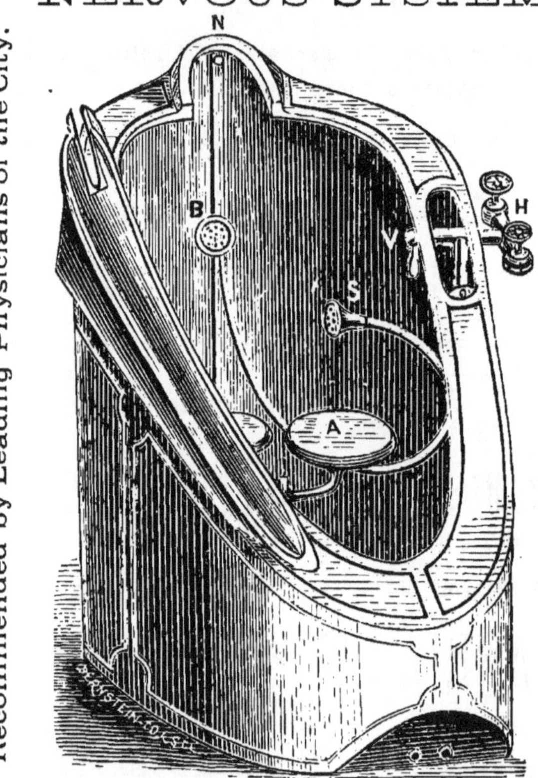

These Statements are made after an experience of years in its use in hundreds of cases

It is pre-eminently useful in the prevention and cure of all

DISEASES PECULIAR TO THE FEMALE SEX.

As a ***BEAUTIFIER*** *of the* ***COMPLEXION*** *it is unrivaled. As a remedy in Dropsy it is equal in efficacy in effecting a cure to all medicines and other agents known to science.*

MR. and MRS. S. F. HINCKLEY,

62 S. Elizabeth St.

GEORGE H. GALE, Manager. CHAS. M. TUBBS, Agent.

CHICAGO CARRIAGE AND CART CO.

WHOLESALE AND RETAIL DEALERS IN

CARRIAGES, BUGGIES, CARTS, CUTTERS,

HARNESS, ROBES, ETC.

356 Wabash Avenue, Chicago.

Good Horses for sale or exchange at all times.

DOCTORS' PHAETONS AND PONY RIGS A SPECIALTY.

Can Undersell any House in the City.

Halliday's Blood Purifier,

THE GREATEST BLOOD PURIFIER ON EARTH.

Rivenburgh's Golden Era,

A PURELY VEGETABLE POWDERED COMPOUND.

Rivenburgh's Brain and Nerve Tonic.

THE SUCCESS OF THE NINETEENTH CENTURY.

The exclusive agency for the above well-known remedies held by

MRS. P. J. L. FINLEY, 237 W. MADISON ST.

THE LUNDGREN INSTITUTE

FOR

Massage and Movement Cure.

LEONARD LUNDGREN, M. D., } Proprietors.
S. A. LUNDGREN, M. D., }

OFFICE: 29 & 30 CENTRAL MUSIC HALL.

CHICAGO, ILL.

OFFICE HOURS: 10.00 to 12.00 a. m.; 3.00 to 5.30 p. m.

CHRONIC DISEASES A SPECIALTY.

THE LAKESIDE HOSPITAL,

3545 Vincennes Avenue, CHICAGO,

NEAR COTTAGE GROVE AVE. AND 35TH ST.

Is a private Hospital, equipped with all modern conveniences, exclusively for the treatment of the

DISEASES OF THE RECTUM AND THEIR COMPLICATIONS

The Institution and all its facilities are at the service of the Profession. For information call on or address

DR. EDWIN H. DORLAND,

9 to 11 at the Hospital; 2 to 4 at City Office, room 516, Chicago Opera House Building. Otherwise, at residence, 4329 Lake Avenue.

TELEPHONE 9982.

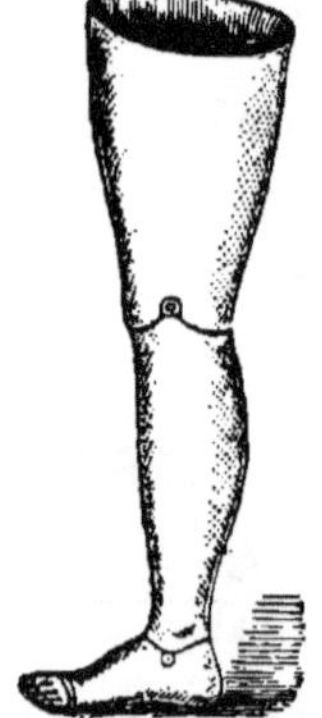

W. H. MONTGOMERY,

MANUFACTURER OF

ARTIFICIAL * LIMBS,

NO. 73 CLARK ST.,

Corner Randolph.

ROOMS 5 AND 6.

FOLKE-KJELLBERG,

OCEAN BRINE, ELECTRIC, MEDICATED AND

VAPOR BATHS,

CENTRAL MUSIC HALL,

(Rooms 47 to 51.)

Genuine Swedish Movements and Massage Treatments administered by trained and graduated Manipulators from the old Country.

Finely appointed apartments for the accommodation of South-Side patrons at 365 WABASH AVENUE, with equally skilled attendants in charge. The whole under the management and immediate supervision of TEKLA FOLKE-KJELLBERG, 365 WABASH AVENUE and CENTRAL MUSIC HALL.

REFERENCES—The Medical Profession at large.

CHICAGO GYNECOLOGICAL INSTITUTE

FOR THE TREATMENT OF THE

Medical and Surgical Diseases of Women.

No. 5306 JEFFERSON AVE., HYDE PARK.

DR. LUCY WAITE. DR. CLARA W. PEASLEE.

Refer, by permission, to the Faculty of Hahnemann Medical College and Hospital of Chicago.

This Institution is for the treatment of Diseases of Women exclusively. The best surgeons in Chicago operate in cases requiring surgical interference. Chronic cases which cannot be treated properly or successfully at the home of the patient, are especially solicited. Physicians placing their patients in the Institute are kept informed in regard to their condition, and are always welcome in consultation. Trained nurses, only, are employed, and the best of care guaranteed. For further information, address

DR. LUCY WAITE,

5306 JEFFERSON AVENUE, HYDE PARK, CHICAGO. Telephone No. 9859.

SPECIALTIES:

Physicians', Dental and Surgical

RUBBER GOODS.

GEO. E. PRESTON,

181 State Street,

MASSAGE

AND

Magnetic Treatment.

Experienced Operators.

M. A. TOWNE,

530 W. Madison St.

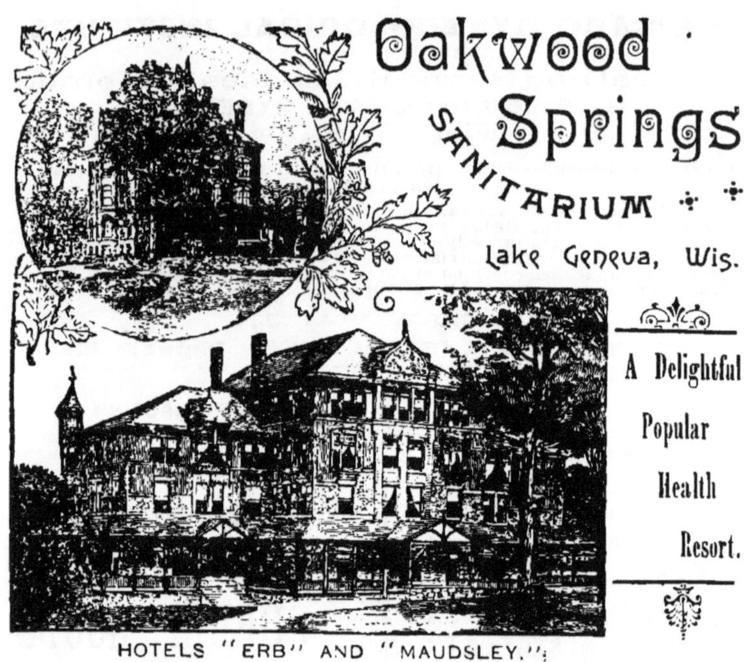

HOTELS "ERB" AND "MAUDSLEY."

A Charming Summer Retreat

AND A

Perfect Home of Winter Comfort.

This Sanitarium is conducted exclusively for the

TREATMENT of DISEASES of the BRAIN and NERVOUS SYSTEM,

including Nervous Diseases of Children, Impediments of Speech, Nervous Prostration, Motor and Sensory Affections of the Nervous System, Neurasthenia from Toxic Agents, Mild cases of Mental Disease, and Gynecological cases when complicated by Nervous Derangement.

Hotels widely separated from each other with independent grounds; perfect classification—no annoyances from other patients; elegant Apartments, Private Parlors and Dining Rooms; Competent Physicians; Skilled Nurses and Attendants; Massage, Electricity, Turkish and Russian Baths. Please see pages 121, 137, 170, 183, 190, and for further information address

Oscar A. King, M. D., Supt., 70 Monroe St., Chicago,

THE

Medical and Surgical Directory

OF

Cook County, Illinois.

1888-89.

Physicians,
Surgeons,
Dentists,
Druggists.

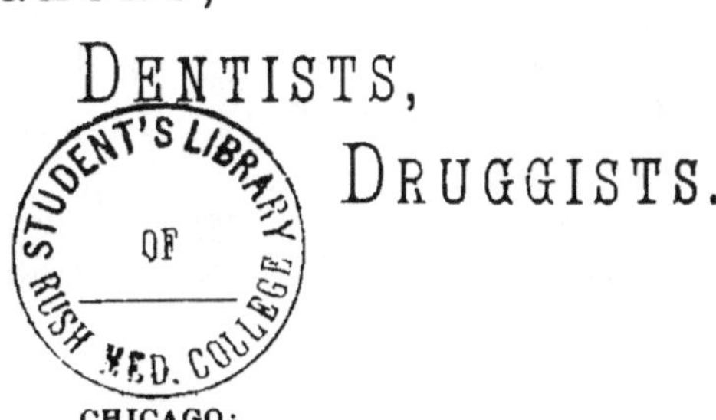
Student's Library of Rush Med. College

CHICAGO:
Physicians' Collection Association
115 Dearborn Street.

CONTENTS.

FEB 17 1900
3947

INDEX TO ADVERTISEMENTS.

JAS. I. LYONS,

Experienced Manufacturer of

ARTIFICIAL LIMBS.

78 Fifth Avenue, Room 3,

CHICAGO, ILL.

A Perfect Fit and Satisfaction Guaranteed.

· · THE · ·

ROSE HILL CEMETERY COMPANY.

OFFICERS:

FREDERICK TUTTLE, President.

VAN H. HIGGINS, Secretary.

E. C. LONG, Treasurer.

City Office, 157 to 163 LaSalle Street.

☞Telephone Directions: Ask for City Office, No. 615; Cemetery Office, No. 3901.

Chicago Physicians' Street Guide.

3347

Aberdeen.

51 C I Thatcher

Adams.

103 C H McCallister
109 P F T Ehlers
109 E A Bassett
113 Eugene S Atwood
113 Fred Roesch

Adams W.

320 J H Plecker
333 G S Thomas
384 John W Hutchins
434 Harold Moyer
470 James H Gates
507 E Fletcher Ingals
530 Ed L Holmes
533 Henry M Lyman
583 Wm M Tomlinson
590 T C Duncan
596 Mrs E H Root
682 Frank W Reilly
683 Glenn M Hammon
686 E Schottenfels
692 Sarah A Anderson
705 Charles Gilbert Davis
719 Seth S Bishop
828 A E Baldwin
844 H A Phillips
878 B W Rogers
885 N Derexa Morey
1043 Lucius O Gibbs

Aldine sq.

20 Charles F Bassett

Allport ave.

719 M Lorenz

Archer ave.

2203 E Lackner
2204 G W Goodner
2204 Willlis F Moore
Cor Clark, M S McGorran
2495 Daniel H Sullivan
2505 Chas H McGorray
2505 Alex Behrendt
2512 Daniel H Sullivan
2802 Wm Wilson Coker
2871 Jacob Rosenthal
2876 G Steurnagel
2894 Frank R Webb
2899 D Egan
2906 Theodore S Bidwell
2906 Paul E La Barriere
2910 A C Alex
3100 Chas C Sperry
3131 W B Cook
3169 Geo E Willard
Cor 38th, J Lee Mills

Armitage.

193 C D Manning
537 J. Struble.
541 J C Merrick

Ashland ave.

23 Curtis M Beebe
130 A W Woodward
162 A C McChesney
170 P H Conley
47 Edward D Strong
172 Adam Miller
176 George F Hawley
559 S Luther McCreight
804 Frank J Jirka
806 F H Henry
847 G Theobald
Cor 12th, J W Mott

Ashland ave. N.

Cor Milwaukee ave.
Alex Thuemmler
613 Herman Schwuchow

Astor.

85 Edwin J Gardiner

Beethoven Pl.

46 Edmund A Boas
46 Desire Q Scheppers

Belden ave.

318 A Underwood Carr
394 William H Weaver

Bissell.

269 I P Farley
278 James F Graham

Blue Island ave.

99 Henry Venn
107 C E Cyrier
109 Denis Morin
134 Walter W Buchanan
156 Joseph N Cadieux
171 William McCarthy
171 T F O'Malley
172 Denis Morin
173 C E Cyrier
173 J M Hutchinson
208 Frederick E Sherman
249 Geo C Synon
249 W A Synon
303 J Maher
322 G Z Bachelle
324 William Martin
360 J H Krueger
361 A Sterl
378 Homer O Bates
378 C E Greenfield
379 Edelmar D Saint-Cyr
379 Emiline D Saint-Cyr
387 Robert Mitter
544 J W Valpey
547 D Birkhoff
547 W H P Evatt
547 Gustav Schirmer
573 Edmund Christie
573 F J Patera
606 F J Maschek
645 W C Ohlendorf

Butterfield.

3008 D F Brennen
3139 Geo Cleveland Hall

Bryan Pl.

1 John C Bryan

California ave. N.

Cor North ave.
Chas H Carter
908 J R Struble

California ave. S.

1111 William E Miller

Calumet ave.

2220 Chas Gilman Smith
Cor 26th, E E Kerr
Cor 26th, Maurice Scheuer
Cor 26th, W D Storer
2600 Frederick W Mercer
2823 Eugene S Atwood
2920 Ransom Dexter
2972 C E Laning
3014 Benj L Colwell
3201 B Rel Van Doozer
3226 Franklin B Ives
3239 J M Pillsbury
3343 Wm E Casselberry

Campbell ave

193 M B Blouke

Canal S.

503 A P Kadison
505 Ernest Epler
505 George W Webster
594 J W Propeck

Canalport ave.

60 Isador Munzer
134 F J Maschek

136 P H Matthei
136 Wilhelm Thies

Carpenter N.

10 J H Mellinger

Carroll ave.

429 W B Marcusson
517 J F Cook
666 Nathan Breiton
733 Charles J Lewis

Cass.

113 Arnold P Gilmore
190 Omer C Snyder
214 Peter S Arndt

Center ave.

21 J B Bell
106 John H DeWitt
117 James G Sinclair
18 Julius Ulrich
202 R Ulrich
395 Frank Taliaferro
218 F W Fitzgerald
453 W H Morgan
455 John O Hobbs
455 Frank Taliaferro
585 Gustav Fischer
585 Fred R Formaneck
631 F H Henry
631 Frank J Jirka

Center ave. N.

73 H S Borchsenius-Skov
114 N E Remmen
116 B M Behrens

Center.

342 S H Bottomley
360 Laura E Boyd
402 Henry M Hobart
407 E E Gwynne
411 John Teare
411 P M Woodworth
427 John F Williams

Chestnut.

269 Clarence W Leigh
559 G B Bushee

Chicago ave.

62 W P Verity
64 Sven Windrow
68 Arvid H Wimermark
90 E Kopp
111 Claes Wm Johnson
177 Julius H Lee
200 E P Ryan
202 D H Stern
207 A H Gordon
308 John A Benson
323 E B Donoghue
355 R W Steger

Chicago ave. W.

239 O B Howe
322 Theodore Nielsen
467 G M Emrick
468 Julius Otto
518 John Tascher

567 Stephen A Hemmi

Claremont ave.

534 A C Morey

Clark.

41 Edwin Bowell
41 Wm Carleton Hunt
41 N Rosenberg
46 C L Downer
59 S J Avery
59 J R Wilkins
73 A C McChesney
81 J F Cook
81 Charles B Gibson
97 Marshall D Ewell
Cor Washington, Jas N Banks
107 G K Dyas
107 W Godfrey Dyas
112 William T Belfield
112 T J Bluthardt
112 John H Chew
112 P H Cronin
112 E H Dorland
112 E W Edwards
112 G F Fiske
112 N H Henderson
112 Ferdinand Henrotin
112 Charles C Higgins
112 Ed L Holmes
112 P C Jensen
112 J W Koehn
112 U G Latta
112 G Frank Lydston
112 Truman W Miller
112 Wm T Montgomery
112 Edwin Pynchon
112 T J Reed
112 S N Schneider
112 C E Webster
117 R D MacArthur
117 W C MacGillis
125 R Emery
125 Wm Krusemarck
125 O W F Snyder
125 Geo B Walker
125 J G Williamson
128 Ben P Reynolds
133 David W Graham
133 W F Lewis
136 Arthur L Blunt
159 M Phillips
159 F Montrose Weller
167 John Kean
186 Franklin F Clarke
188 H N Small
193 George A Cutler
193 Wm Hanley
193 Sidney Walker
199 Jno F Bigelow
199 Samuel Edwards
199 W D Lonergan
201 Charles Bigelow
207 Sereno W Ingraham
210½ F J Tongue
214 R W Steger
214 W H F Smith
250 A H DePuy

250 Chas W Purdy
250 John H Rauch
277 T L Carey
277 John E Chambers
279 W H Landis
279 A C Lucas
281 J Sumney
329 V Centaro
349 J Harvey Bates
349 T P Shanahan
351 H Kirschtein
351 Peter McDonald
351 Edwin O Wilms
Cor Harrison, H W Jay
418 G Ronga
422 F X Barter
480½ John Bartlett
2011 Louis N Barlow

N. Clark.

26 Carl A Helmuth
26 Gerhard C Paoli
27 H H Cook
60 Joseph Krost
81 W D Garvin
105 Jessie G Forrester
109 Edwin Turnock
146 W Reinhold
152 W G Eggleston
152 Geo E Richards
167 Clement Venn
171 M Gill
181 Lucius O Gibbs
188 Ward Greene Clarke
208 M Gill
237 George Cunningham
307 F G Bonynge
341 W T Thackeray
343 Ph D Paul
353 John D Kales
355 C A Wilbur
445 William T Belfield
448 Karl Herz
456 Lottie J Park
468 P H Cronin
481 H Geiger
486 Rosa H Engert
509 Th Heuchling
825 Thilo Brauns
1071 Truman W Miller

Clarkson Ct.

11 D W Hunt

Cleveland ave.

458 A Roesenbury
548 J Simpson

Clybourn ave.

103 H Kleist
120½ Edmund A Boas
135 H F Ruschhaupt
297 Franz Hoegelsberger
297 Hans Hoegelsberger
858 J J Bell
864 J J Bell
876 J R Brandt

Colorado ave.

223 John T Gray

Congress W.

488 John Edwin Rhodes
577 Thos J Shaw

Cottage Grove ave.

2412 Fred D Rogers
2448 O E Schultz
2448 C P Stringfield
2448 F M Stringfield
2622 Harold Jacobsen
3100 J A McGaughey
3240 Clement Hallowell
3266 Elbert Wing
3411 Julia M Orr
3640 F B Norcom
3640 J Priestman
3701 Arthur W Bigelow
3872 Charles F Bassett
3872 A L Lennard
3900 T S Huffaker
3900 W H Schrader
3901 C H Buchanan
3904 Geo F Parsons
3962 M H Lackersteen

Curtis.

182 J F Stanton

Dayton.

332 Cyrenius A David

Dearborn.

C S Eldridge
69 Henry M Lyman
69 I N Danforth
69 James G Kiernan
71 L B Hayman
71 L B Haynes
110 Andrew James Park
112 W H Burt
130 W Gary Le Roy
130 E W Leroy
168 Arthur Burley Hosmer
184 Odelia Blinn
204 V L Hurlburt
204 Wm F Smith
225 C D Calkins
359 Nicholas Re
444 A Royce Camp
2407 H E Hildebrand
3424 Isaac N Ellsberry
3728 Katherine D Munn
3800 John F Abel
3838 C M Haynes
3838 G F Haynes

Dearborn ave.

102 J E Ryan
114 Lucy R Meyer
131 W A D Montgomery
Cor Erie, Sarah Hackett Stevenson
153 C Storck
166 C D Calkins
181 Leila G Bedell
210 S N Schneider
215 Henry Hooper
220 John H Wilson
234 Alexander H Cooke
234 J Masson Cooke
241 William J Hawkes
261 Robt N Tooker
Union Club, Charles H Vilas
321 George S Isham
321 Ralph N Isham
414 R D MacArthur
432 C O Rutter
441 L D Rogers
441 S Ida Rogers
487 Samuel J. Jones

Deering.

3014 Jacob Rosenthal

Desplaines S.

138 Peter J Rowan

Division E.

229 Arthur R. Reynolds
247 F A Hess
276 Alexander Thuemmler
305 J C Hoag
307 Hugh T Patrick
318 F B Eisen Bockius
360 H. H. Look
360 W J Webb
374 F P Devries
399 Augusta Hinz
607 F L Wadsworth

Division W.

177 James K Bartholomew
296 S A Hemmi
736 Chas E Cessna

Eighteenth.

68 J E Stubbs
165 William H Doolittle

Eighteenth W.

731 Frank J Novak

Elizabeth S.

43 U J Swain
47 Geo W Wolgamott
57 Austin A Wescott
56 Cassius D Wescott

Elk Grove ave.

789 F L Breidenstein
846 Fanny Leake

Ellis ave.

3523 F S Tabor
3709 Mrs C E Weilhart
3711 L Bedford
3714 F R Wright

Ellis Park.

3621 John G Trine

Elm.

nr. Market, P E Torgney Anderson

Elston ave.

721 L S Bowers
721 D G Moore

Erie.

200 John Louis Irwin
237 G Hessert
269 Robert H Babcock
281 Nicholas Jas Dorsey

Erie W.

192 Jennie E Hayner

Eugenie.

23 Carl Meyers Mentzel
174 B L Heegard

Evergreen ave.

34 N Rosenberg
142 Jacob Dal

Ewing Pl.

20 Chas H Carter

Fifth Ave.

95 Henry Dietrich
347 Chris Fuersattel

Fillmore.

1161 Nelson H. Church

Forest ave.

3100 Henry T Byford
3226 Jas G Kiernan
3245 H B Sanders
3245 W H Sanders.
3301 Marcus P. Hatfield.
3305 Louis L McArthur
3613 T L Carey

Fourth ave.

147 A Royce Camp

Fowler.

52 Niles T Quales
60 Baltazar Meyer

Franklin N.

178 Harry S Buffum
384 Lafayette W Case
663 E J Brougham

Frederick.

1834 Gerhard C Paoli

Fremont.

52 Adolph Buttner
220 Christian Fessel

Fullerton ave.

73 Geo B Connolly
82 Friedrich Bersuch
102 C Rutherford
124 Edwin O Wilms
136 C F Nitz
490 Valentine A Boyer
513 J E Hequembourg
570 G Frank Lydston
576 S C Blake

Fulton.

315 Saul J Holmes
341 Charles W Evans
629 W P Phelon
667 Allen W Gray
667 Ethan A Gray
878 J S Hunt

Gilpin Pl.

18 G Z Bachelle

Grant Pl.

28 Anna M Parker
64 E L Rivenburg

Graves Pl.

3212 F M Madison
3214 Edwin C Williams

Grenshaw.

1085 Charles M Clark

Gross ave.

854 Robert Bradbury

Groveland ave.

2924 Joseph Zeisler
3033 B C Miller
3166 J Matterson
3234 C K Fleming

Halsted.

9 J R McCullough
Cor Washington, R W Jones
62 O T Shenick
100 Henry C Feder
102 Gertrude S Condit
126 Cornelius Hogan
133 Siremba Shaw
134 Stuart Johnstone
168 W C Caldwell
168 J R Bedford
208 William A Howard
232 J S Lane
234 Richard Lull
246 Ph H Matthei
254 C F Moore
254 T A Clark
260 C E Canfield
262 Edward W Lee
277 A S McLennan
262 F S Hartman
262 John B Murphy
262 George E DeMarr
323 Frank W Stebalts
Cor Van Buren, Oscar J Price
Cor Polk, F M Tebbetts
Cor Polk, Geo S Gfroerer
376 J J McGrath
527 Homer O Bates
527 Th Rahlfs
551 Louis Braun
564 J J Thometz
571 Emil Schottenfels
583 J L Mulfinger
725 Wm King
729 Frank W Stebalts
737 E G H Miessler
745 Wm King
750 G Theobald
802 Homer O Bates
802 Stephen C DeVerry
802 Laura Cowgill-Bates
802 Nelson H Church
802 J H Curtis
Cor 14th, Chr Schleyer
2603 Louis Braun
3101 W J C Casely
3452 P G Jennings
3552 Bruce Miller
3659 James G Berry
3701 J R McNamara
3718 James Burry
3835 Joseph Reilly

Halsted N.

182 C E Brinckerhoff
657 J A Heckenbach
800 C F Nitz
954 Chas C Bernard
Webster ave, C S Bacon
1102 E Ilee Kerlin
1130 Edward Everett
1594 Fred D Porter

Harrison.

98 S B L Merrill

Harrison W.

160 Peter J Rowan
339 D P Hueston
395 E A Fillmore
395 J F Stanton
440 J H Curtis
Cor Center ave, W B Marcusson
583 George L Beach
647 D Birkhoff
863 R S Hall
941 L K Blakeslee
947 Joseph Shugart
951 Laban S Major
1001 Guy B Dickson
1010 Ellen H Heise
1070 R C Knox
1084 D Duncan
1123 Alexander M Stout
1150 A C Bradway
1190 J E Roop

Hermitage ave.

251 J J Thompson

S. Homan ave.

923 Geo R Bassett
923 Susan A Bassett

Homer.

62 Patrick Curran

Honore.

278 J B Brooker

Howe.

175 E M Landis

Hoyne ave.

113 Charles B Gibson
155 J R Wilkins
237 Joseph M Patton
310 James P Prestley
355 J J Schnbert

Hoyne ave. N.

643 D G Moore

Huron.

272 R L Rea
291 N S Davis Sr
291 John D Kales
229½ Ambrose Breese

Illinois.

235 W H Marble

Indiana Ave.

1434 T B Swartz
1424 E H Schwandt
1619 E C Dudley
1800 E Russell Ogden
1816 Elmer E Babcock
1822 M Mannheimer
1832 William H Byford
1833 Temple S Hoyne
1931 Mary A Mixer
1922 George W Webster
2306 Thomas Telfer Oliver
2330 William W Jaggard
2400 Daniel T Nelson
2500 Geo H Weeks
2503 J B Pettengill
2510 Milton Jay
2542 Stephen C DeVerry
2500 Mary Weeks Burnett
2548 Almon Brooks
2719 Henry L Deimel
2829 Peter McDonald
2920 D A K Steele
2958 J W Brooks
2969 H B Fellows
3018 Ebenezer H Thurston
3025 Henry W Boyd
3027 Frank Cary
3027 Harriet H Cary
3031 Geo M Chamberlin
3130 Geo F Shears
3136 Charles H Lodor
3151 E D Cross
Cor Thirty-Second st William E Quine
3237 Chas Caldwell
3302 Horace M Starkey
3400 John Leeming
3422 F A Emmons
3423 W H Davis
3449 Frank E Waxham
454 J R Kippax
3546 E A Bassett
3619 Henry Olin
3638 Jos P Cobb
3670 U G Latta
3837 J Sumney
3904 N C Kemp
3926 Finley E Ellingwood

Indiana.

182 D H Stern
297 Lewis H Watson

Indiana W.

199 F B E Bockius
210 S Douglas Twining

210 Randolph N Hall
223 Carl Sandberg
280 Henry Bell
282 Merritt W Thompson
284 A Doe
312 R J Piper
312 Albert B Strong
477 J B Phelan
528 William Deegan
528 E P Koch
530 William H Hayman
627 J R Buchan
676 Randolph N Hall
729 H H Latimer
864 Chas J McIntyre

Ingraham.

50 M P Kossakowski

Jackson.

Argyle bldg
Samuel J Jones
Cor Michigan
John S Marshall
Argyle bldg
H M Scudder
126 B D Coop

Jackson W.

254 W H Landis
362 Theodore Bidwell
365 M W Borland
365 Leonard C Borland
432 John O Hobbs
454 W H Morgan
547 J M Hutchinson
570 Norman Bridge
597 Daniel R Brower
625 Oscar J Price
638 Mary H Thompson
723 Jno F Bigelow
778 J B McGinley
820 John C Webster
842 F Brooks
880 Elizabeth H Trout
942 J L Mulfinger
1214 Luke A Harcourt
1237 George F Yates

Jefferson.

339 Alexander M Stout

Johnson ave.

84 Wm Robinson

Johnson Pl.

3805 H O Rockwell
3834 W H Schrader

Kinzie.

47 W W Colton
Cor Kinzie and Market
Wm H Eldred
56 J D Andrews

Laflin.

16 Alfred Hinde
30 Mary H Bowen
101 Edward W Lee

Lake W.

206 A K Smith
482 J J M Angear
537 L W Whitney
626 E M Smith
628 Harry L. Barnum
628 Ira E Marshall
674 J Brown Loring
696 A G Uaerther
722 Allen W Gray
894 J Warren Walker
966 A H Tagert
966 Frank M Ingalls
966 Alonzo D. Tagert
979 Wells Andrews
1001 M H McGrath
1012 A W Smith
1333 R H Kenning

Lake ave.

3912 N B Delamater.

Langley ave.

3748 Edgar M Reading
3750 Edgar Reading

Larrabee.

292 Desire Q Scheppers
225 J Saalfeldt
358 M G Kellner
405 H Rakenius
681 Charles M Downs
802 D. Rosenthal

La Salle.

114 Wm Robinson
132 A H Foster
127 Robert Greer
152 Levi R Jerome
161 Alice B Stockham
187 J R Gore
205 Geo O Taylor
218 Th Heuchling
330 Charles Pusheck
3249 A W Collier
3616 Bruce Miller
3726 Albert E Froom

La Salle ave.

177 Henry Le Caron
229 C E Manierre
269 Christian Fenger
Cor Locust Ward Greene Calrke
344 Bernard S Arnulphy
285 J R Boynton
353 Ferdinand Henrotin
380 Chas White
396 J W Niles
410 Geo C Hunt
418 James P Buck
427 W E Reed
435 Halleck Hart Look
438 G F Fiske
453 T J Bluthardt
456 Henry Banga
499 Charles M Downs
519 E H Pratt
545 Fred Roesch

655 N P Pearson
558 J S Clark

Leavitt.

479 William Rittenhouse
547 Vira A Rockway
677 O Wegner

Lincoln.

193 Frank H Montgomery
215 Joseph Shugart

Lincoln N.

296 W J Neill

Lincoln ave.

17 J Frank
18 Chas Pague
51 Lawrence H Prince
51 Charles T Parkes
78 Lawrence H Prince
84 E G Trimble
138 A L Farr
Cor Garfield Blanche Moore
163 Albert Goldspohn
185 Lemuel C Grosvenor
239 Arthur G Thome
244 Corresta T Canfield
245 Frederick S Senier
245 Shobal Vail Clevenger
315 Isadore L Green

Lock.

3015 J R Richardson

Loomis.

14 J Suydam Knox
90 Henry Plumb
116 Rufus H Bartlett
132 J W McCausland
148 John E Harper
109 J H Greer
148 E P Murdock
156 K Schaefer
209 W W Sheppard
274 William Y Provost
285 B D Colby
310 R W Jones

Lowe ave.

120 G Steurnagel

Lytle.

62 John Fellbrich

Madison.

48 James E Gross
70 F R Sherwood
70 F S Tabor
76 Ransom Dexter
78 Junius M Hall
78 J E Ryan
80 J Harvey Bates
82 Emil Sincere
89 Rufus H Bartlett
113 Beach Loomis
119 Geo B Abbott
119 W Harrison Ballard
119 J B McGinley

119 G J White
171 Henry Tomboeken
196 Archie J. Shimp
197 George W Newton

Madison W.

50 H A Phillips
82 A G Mauro
82 Charles F Perkins
82 Jos R Waldmeyer
115 A B Bausman
161 E P Murdock
161 W W Sheppard
169 J J Davis
181 John T Gray
181 A J C Saunier
182 R L Mintie
183 J H Plecker
186 T T Taylor
231 J B Bell
247 C Jackson Tobias
247 R B Treat
253 Carlos J Adams
255 S W Howe
257 Julia R Howe
259 Maranda E Hyde
292 Mary Shaffer
321 D F Boughton
325 Arthur B Freeman
325 A E Kauffman
424 J R Bedford
437 J B Armstrong
454 W Z Flower
471 James K Egbert
490 John A McDonnell
552 L J Davis
557 E C Sweet
561 Edwin Pynchon
565 W M W Davidson
565 W S Harvey
567 Cassius D Wescott
688 F J Dewey
729 Louisa Martin
737 Philip Adolphus
828 Elmore S Pettyjohn
884 Henry A Fitch
884 Weller Van Hook
886 H A White
904 Alferd H Stephani
937 Bryon Wilson Griffin
1001 Vincent Haight
1001 Charles Davidson
1251 A F Snyder
1251 George White

Main.

2834 Mary J Kearsly

Market N.

279 F M Constant
378 Anton Lakey
539 Julius W Oswald

Marshfield ave.

222 N B Rice

May N.

15 Henry Van Buren
49 J M Page
265 Snsan E Bruce

273 William M Wilke
279 Sarah J Hogan

Michigan ave.

Cor Adams,
Melville A Tully
Cor Jackson,
J D Hammond
192 Mrs Prudena Saur
194 Wallace C Clark
271 A Reeves Jackson
348 H M Hunt
348 J C McCoy
348 H G Wildman
1255 David S Smith
1436 R Tilley
1528 H Merckle
1634 James H Etheridge
1636 E J Ogden
1823 Reuben Ludlam
1806 John E Owens
1823 Reuben Ludlam Jr
1823 Mrs Belle L Reynolds
2001 J Adams Allen
2123 Elmer L Hollister
2202 E L McAoliff
2211 Georgia S Ruggles
2112 J B Talcott
2308 Emma F Gaston
2239 Henry P Merriman
2353 William T Akins
2490 James N Hyde
2415 Jas N Banks
3002 G K Dyas
3030 Curtis T Fenn
3030 J P Morrison
3034 R C Bain
3034 E S Baily
3034 D H Williams
3125 Edwin J Kuh
3237 Alex Loew
3456 S A McWilliams
3907 Frank R Webb

Millard ave.

958 Charles Amet

Milwaukee ave.

126 Wm O'Connell
126 Charles W Evans
126 W J Neill
208 B A Brigham
212 J C Pickard
250 M Connor Moran
236 Rocco Brindisi
232 G Warren Reynolds
234 F B Eisen Bockius
241 Baltazar Meyer
241 Niles T Quales
246 J Julson
257 Sigismund D Jacobson
318 F Kleene
428 John A Brill
449 C Giegerich
465 Theo Wild
469 Frederick Schaefer
476 D W Nolan
480 D C Stillians
482 M L Harris
493 G Thilo

496 Edward Mead
547 Chas H Venn
570 F Hasse
625 F W Boeber
707 Adolph Rosenthal
783 Richard Westerburg
829 Julius Buzik
845 J K Bartholomew
845 Louis G Koier
845 Charles M Koier
845 John Tascher
860 J W Koehn
860 John W Dal
875 O Wegner
709 F J Patera
994 Carl Hattermann
996 Abbie A Hinkle
Cor Lincoln,
Geo Leininger
1020 Gustav Fernitz
1218 Henry J Burwash
1228 John Bergeson
1475 William Bein
1546 Chas E Ceesna
15?3 Thos M Gandey
1602 G O Shettler
1606 Thomas Jesperson
1906 G B Bowlby

Mohawk.

322 L P Griggs

Monroe.

Cor Wabash, Clifton house,
Julia C Whaling
34 J Blanchard
34 Wallace Blanchard
70 William L Axford
70 Robert H Babcock
70 Charles H Beard
70 F G Bonynge
70 Wm E Casselberry
70 E C Dudley
70 Edwin J Gardiner
70 Arnold P Gilmore
70 J Lucius Gray
70 John H Hollister
70 Arthur B Hosmer
70 George Isham
70 Ralph N Isham
70 Oscar A King
70 J Suydam Knox
70 E J Ogden
70 E Russell Ogden
70 Frank E Waxham
70 T B Wiggin

Monroe W.

285 L C Fritts
289 Thaddeus P Seeley
294 I N Danforth
296 T Davis Fitch
334 Andrew J Baxter
343 Mrs H E Stansbury
372 L B Haynes
397 E M P Ludlam
419 Sydney Walker
489 J Elliott Colburn
522 W T Nichols
539 Leonard St John

552 Albert G Beebe
533 Albert B Strong
554 Henry P Newman
595 Stuart Johnstone
619 W F Lewis
645 Calvin M Fitch
645 Walter M Fitch
651 William De Bey
651 Ida Mueller
672 C W Hempstead
980 S K Crawford
690 William E Clarke
708 E Marguerat
779 A H Foster
805 Frank M Ingalls
850 Adolph Rosenthal
856 Carlos J Adams
866 W L Copeland
903 James P Mills
913 C D Messinger
913 E D Messinger
963 Hugh P Skiles
1020 C L Downer
1028 S K Falls
1205 A F Snyder
1549 A W Smith

Moore.

145 -Mathias Brand

Morgan.

46 D B Deal
90 F A Stanley
109 Ferdinand C Hotz

Morgan W.

196 A G Mauro

Noble.

285 W D Neel
479 H Steinhouse
562 Moriah Orgleit
615 C Revkowski
743 Wenzel Majeski

North ave.

150 G M Illingsworth
159 J W Hendricks
159 G M Illingsworth
159 E Iles Kerlin
191 F Scheuermann
232 J W Hendricks
240 Charles Koller
255 Samuel Schaefer
288 H Rakenius
302 V Schreiner
315 W W Wetherla
451 J H Hoelscher
454 Paul Kreve

W. North ave.

237 Patrick Curran
239 L S Bowers
256 M P Kossakowski
498 Henry H Sloan
570 Charles H Evans
575 Louis G Koier
587 M H Luken
804 L E Lawson
915 Joseph H S Johnson

Oak.

104 C E Webster
128 P E T Anderson
152 J Simpson
156 C E Manierre

S. Oakley ave.

227 J J Tuthill
307 Robert Greer
402 Frank H Booth
403 M M Thompson
429 J E Reynolds
433 Charles S Taylor
631 Francis C Caldwell

Nineteenth St. W.

56 Friedrich B Meyer

Oakwood boul.

144 R D Boyd
144 Milton F Coe
144 H H Deming
144 Alice Ewing
144 H O Rockwell
144 J S Stockwell
1316 Martha A Bowerman
1339 Geo O Taylor
1421 J C David

Ogden ave.

98 C C P Silva
230 Stella B Nichols
245 F Brooke
246 E W Le Roy
358 Charles Amet
358 J E White
358 J B Williams
358 Carrie Noble White
447 George E DeMarr
481 Charles S Taylor
987 William Rittenhouse
1324 Geo R Bassett
1324 Susan A Bassett
1536 French Moore

Ohio.

205 J K Winer
264 James T Sharpe

Ohio W.

583 H B Fenner

Ontario.

277 C F Ely
366 Joseph H Buffum

Orchard.

252 L V Kalkstein

Page.

85 D W McNeal

Park ave.

22 W R Griswold
70 James Bruelly
70 James Bradley
118 L B Montgomery
118 Liston Montgomery
138 John E Hurlbut
171 J F Hopkins
182 J W Meek
243 Thomas D Palmer
274 O B Damon
321 Chas E Walker
321 J D Walker
329 Joseph S Sageser
368 G H Carder
412 Eugene E Holroyd

Park Row.

5 John Blair King

Paulina S.

3243 Chas C Sperry

Paulina N.

749 Christopher Traub

Peoria N.

150 Wm O'Connell

Pine.

47½ Charlotte E Frink

Polk W.

873 I W Brown
977 William A Howard

Portland ave.

2727 Otto Poppe

Prairie ave.

1620 Lyman Ware
2001 John W Streeter
2011 De Laski Miller
2228 E W Hunter
2237 Staley N Chapin
2332 Horatio N Hurlbut
2400 H V Halbert
2400 George A Hall
2406 E J Doering
2447 James F Todd
2606 J R Gore
2954 Charles Gatchell
2954 J S Mitchell
3030 Hel n M Hannah
3343 John S Marshall
3408 J B McFatrich
3543 Horatio Keeler

Randolph.

45 John G Trine
65 Edmund Andrews
65 E Wyllys Andrews
65 Frank T Andrews
65 Andrew J Baxter
65 B A Brigham
65 N S Davis, Sr
65 N S Davis, Jr
65 W G Eggleston
65 James H Etheridge
65 Henry P Newman
65 Edward B Weston
68 Ddward S McLeod
68 L R Williams
89 W E Dodd
89 E Marguerat
89 N P Pearson
89 A Rosenbury
171 Rudolph Seiffert

189 Liston H Montgomery

Randolph W.

56 A C Brendecke
156 Alfred H Stephani
156 Wilhelm Thies
193 J Schaller
199 E Seaman Akely
199 G Kernahan
231 John H Byrne
262 U J Swain
229 John Ring
389 M F Irwin
430 William Y Provost

Ray ave.

39 N J Nielsen

Rhodes ave.

3132 W Franklin Coleman
Cor 32d, Louisa Sedgwick
3308 F H Martin
3312 J G Maynard
3440 John H Hollister

Robey S.

145 Beach Loomis

Robey N.

697 Theo Wild

Rush.

37 B D Foster
49 Norval H Pierce
54 Charles P Beaman

Sacramento ave.

131 W R Van Hook

Sangamon S.

78 Wm McGregor
121 G E Brinckerhoff
246½ B D Coop

Sedgwick.

277 F P De Vries
312 Henry Dietrich
313 Henry Banga
316 Charles Koller
557 E J Fischer
582 Geo J Schaller
657 Oscar O Baines
629 L R Williams
665 Jennie E Smith

Seeley ave.

41 Chas Bert Reed

Seminary ave.

178 John F Runnels

Sheffield ave.

236 W W Elliott

Sheldon S.

10 M G McLean

Sixteenth.

4 Frank S Johnson
4 H A Johnson
6 Edmund Andrews
6 Frank T Andrews
6 E Wyllys Andrews
12 J Harvey Bates
44 Clifford Mitchell

South Park ave.

2612 Jessie G Robertson
2708 Alicia A Flanders
2921 H Kirschtein
3035 C K Fleming
3228 Boerne Bettman
3235 Marcus P Hatfield

Stanton ave.

3616 Henry Sherry

State.

56 D Duncan
58 S A McWilliams
58 William E Quine
58 D C Stillians
64 A G Haerther
69 Harriet C B Alexander
69 Peter S Arndt
69 Bernard S Arnulphy
69 Mary E Bates
69 Boerne Bettman
69 M J Bliem
69 Mary Weeks Burnett
69 Mrs Rose W Bryan
69 Francis C Caldwell
69 Shobal Vail Clevenger
69 James D Craig
69 Rosa H Engert
69 Edward Everett
69 Chas Gordon Fuller
69 John E Gilman
69 Henry Gradle
69 William J Hawkes
69 Jennie E Hayner
69 Samuel P Hedges
69 J E Hequembourg
69 Alfred H Hiatt
69 Edwin J Kuh
69 C E Laning
69 Wells LeFevre
69 Leonard Lundgren
69 Sven A Lundgren
69 Abbie J Mace
69 Marie J Mergler
69 C D Messinger
69 E D Messinger
69 Blanche Moore
69 Henrietta K Morris
69 Lottie J Park
69 E A Pratt
69 Leonard H Pratt
69 Marie E Reasner
69 W E Reed
69 Geo E Richards
69 Mrs E H Roots
69 Sarah H Stevenson
69 C I Thacher
69 Mary H Thompson
69 Charles H. Vilas
69 Joseph Watry
69 Geo H Weeks
69 J A Weeks
69 W H Woodbury
70 Henry Banga
70 H H Boulter
70 J R Boynton
70 Stephen Breckinridge
70 Daniel R Brower
70 Harry Brown
70 W G Bryson
70 Staley N Chapin
70 D B Deal
70 A L DeSauchet
70 Fannie Dickinson
70 C S Eldridge
70 H S Hahn
70 E Fletcher Ingals
80 Myron E Lane
70 Halleck Hart Look
70 E M P Ludlam
70 Lewis L McArthur
70 J Matterson
70 Clifford Mitchell
70 Norval H Pierce
70 John Edwin Rhodes
70 J A Robison
70 Joseph P Ross
70 Vida A Saunder
70 Emilie Siegmund
70 E C Sweet
70 Mary L Vincent
70 W. J Webb
70 Catharine J Wells
70 Geo W Wolgamott
75 I Clendenen
78 M Mannheimer
78 F L Peire
78 A G Schloesser
78 Ernest Schmidt
78 O L Schmidt
96 J G Carroll
100 Joseph H Buffum
100 T C Duncan
100 W M Stearns
103 Joseph G Bemis
103 Charles H Berry
103 H A Coulter
103 E D Cross
103 J C David
103 C F Ely
103 F H Foster
103 Albert S Gray
103 Wallace K Harrison
103 Ferdinand C Hotz
103 E W Hunter
103 Robert Hunter
103 Mahlon Hutchinson
103 M Gaylord Pingree
103 Edgar M Reading
103 Edwin F Rush
103 Fred C Schaefer
103 C A Williams
106 James Heffernan
125 J Allen Adams
125 Rufus W Bishop
125 S C Blake
125 G E Brinckerhoff
125 Henry T Byford
125 William H Byford

125 Cyrenius A David
125 N B Delamater
125 Christian Fenger
125 E L Guffin
125 George F Hawley
125 Bayard Holmes
125 John W Hutchins
125 Franklin B Ives
125 Daniel T Nelson
125 Julius Otto
125 Charles T Parkes
125 James P Prestley
125 Charles G Smith
125 R Tilley
125 Lyman Ware
125 Joseph Zeisler
126 B M Behrens
126 Moreau R Brown
126 J Elliott Colburn
126 W H Davis
126 F J Dewey
126 Frank H Gardiner
126 Milton Jay
126 J B McFatrich
106 J E Stubbs
163 Geo B Abbot
163 Charles P Beaman
65 N S Davis
163 W J Clary
163 W Franklin Coleman
163 W L Copeland
163 Wm H German
163 John E Harper
163 Horatio Keeler
163 M H Lackersteen
163 F H Martin
163 M D Ogden
163 Henry Olin
163 Chas W Purdy
163 Henry J Reynolds
163 C C P Silva
163 W T Thackeray
171 F L Peire
175 M D Ogden
175 Henry J Reynolds
202 A C Rankin
218 Alfred L Willard
224 H G Wildman
224 J C McCoy
235 Frank Billings
235 H H Frothingham
235 Chas Pague
235 Heman Spalding
235 Swayne Wickersham
236 Arthur De Bansset
239 Wallace Blanchard
243 James M Brydon
243 Walter Hay
243 Edwin H Hayes
243 Katherine D Munn
243 Henry K Stratford
243 E B Taylor
279 Albert H Burr
279 C Allison Foulks
359 H W Joy
361 Chas C Singley
363 C Graham
426 H E Hildebrand
424 C H McCallister
452 W L Simpson
441 D Mills Tucker
527 John B Chaffee
511 Anson L Clark
533 Geo Cleveland Hall
511 M G Hart
511 R A Hasbrouck
513 H S Tucker
581 F A Wheeler
511 H K Whitford
1601 Adolphus D Colton
1735 Frank R Walters
1801 Charles C Beery
1801 Elmer E Babcock
1801 T K Jacobs
1801 D A K Steele
2113 C W Stanley
2524 Chas N Durselen
2719 Geo W Miller
2724 Anson Munsell
2725 David B Eaton
2840 Chas J Creighton
2952 A A Wesley
3021 Harry C Whiting
3102 Andrew J Coey
3102 J Guerin
3102 W K Jaques
3102 Edw V McDonald
3102 A A Wesley
3160 John H Besharian
3160 William E Morgan
3300 A R Small
3500 Chas C Fredigke
3504 William E Hall
3656 Finley Ellingwood
3732 W H Wolford
3818 J A Shepstone
3835 Willis F Moore
3847 A J Marks
3858 Charles Caldwell
3901 W H Edgar
3904 J W Marley
3904 W H Wolford
3906 N N Hurst

State N.

59 W L Marr
153 Otto T Freer
177 J Blanchard
177 Wallace Blanchard
205 A Melville Tully
295 Wallace K Harrison
314 Moreau R Brown
384 William C Hunt
417 John H Chew
426 Hugh T Patrick

Superior.

326 N S Davis, Jr

Thirty-First.

52 William L Axford
52 Ed Bert
52 Chas A Storey
52 Geo A Thompson
79 Geo A Thompson
79 T B Wiggin
204 Scott Helm
207 Jos P Cobb
207 Charles A Dewey
217 Vida A Saunder
217 Catharine J Wells
221 J Harvey Lyon
243 F Montrose Weller
268 H B Sanders
300 W M Burbank
300 W J Johnson
339 A J Marks
603 Emil Giljohann
628 W J C Casley
728 A D Finncane

Thirty-Second.

Cor Rhoads ave Rhoda M Galloway
284 C J Simone

Thirty-Seventh.

97 E A Ballard
148 Sheldon Leavett
356 F R Sherwood
358 K L Hickox
445 Chas C Fredigke
540 Arthur L Blunt

Thirty-Third.

118 W H Lewis

Throop.

34 Ephraim Ingalls
34 Homer M Thomas
36 John B Murphy

Twelfth W.

136 Gustav Fischer
136 Fred R Formaneck
Cor Jefferson, F J Jirka
182 A C Morey
287 Geo F Bradley
287 Michael W Kelleher
287 Ed Liebig
287 Frank J Novak
289 John J Alderson
336 E J Howe
336 J E Reynolds
336 James P Way
422 I N Lilly
422 T A Lilly
422 Phil Sattler
492 J H Fitzgerald
509 Chas L Kerler
561 E F Buecking
503 W E J Michelet
Cor Throop, S B L Merrill
621 Carl Struh
659 R Melms
781 Chris Fuersattel
783 T A Scott
908 James Burke
999 J J Thometz

Twenty-Eighth.

205 Charles Adams

Twenty-Fifth.

181 Charles C Beery
189 T K Jacobs

Twenty-First.

122 Louise Dickerson.
642 J R Hayes
681 Edmund Christie.
709 Mathias Brand
709 David O'Shea
722 E A Mullan
730 W H P Evatt
890 Jno G Franke
900 Jennie B Clark
900 William E Miller
900 J A Clark

Twenty-Ninth

34 William Edward Morgan
75½ R H Broe
81 J Young
139 E O Beebe

Twenty-Second.

65 Albert B Hale
65 Edwin M Hale
79½ J B Talcott
125 Chas Krusemarck
125 E O F Roller
125 Horace M Starkey
129 E Dilliard
144 Mary L Vincent
171 Sara E Bacon
171 Elmer L Hollister
171 James F Todd

Twenty-Second W

871 W M Jackson

Twenty-Seventh.

559 N J Schroeder

Twenty-Sixth.

37 Frederick W Mercer
348 D F Brennen
420 Augustus V Park
449 D Collins
477 J A Schmidt

Twenty-Third.

64 S M French
65 Frank Billings

University Pl.

35 Lester Curtis

Van Buren.

112 A C Olin
134 Lysander Meeker
171 E L Stahl, Jr

Van Buren W.

Cor Halsted
Eugene W Whitney
328 Charles Earll
329 H N Small
329 Andrew Stewart
378 Leonard C Borland
378 M W Borland
426 F A Hodson
507 H M Martin
527 Mary E Bates
592 Wm T Montgomery
600 Wm Hanley
694 E L Tons
858 J J Tuthill
971 George W Newton
977 T A Davis
1035 W T Sherwood
1215 G H Scroggy

Vernon ave.

3001 Chas A Storey
3120 Joseph L Hancock
3145 D J Harris
3200 Andrew J Park
3225 Edward B Weston
3229 William T Barry
3242 Ed Bert
3305 Samnel Cole
3401 J Priestman
3600 W R Hillegas

Vernon Park Pl.

36½ Delvina Dion

Vincennes ave.

3545 W V Coffin
3560 J S Stockwell
3602 H H Deming
3629 Kate I Graves
3638 F B Norcomb
3733 Edward W Sawyer
3857 O H E Clarke
3857 E A Hanna
3846 A L Lennard

Wabash ave.

130 Charles G Plummer
143 L D McIntosh
143 T W Sheardown
240 J H Bird
240 G B Bowlby
240 Almon Brooks
240 A C Cotton
240 Charles G Davis
240 Delvina Dion
240 J D Hammond
240 Justin Hayes
240 Plymmon Hayes
240 James N Hyde
240 F H Montgomery
240 D W McNeal
240 Henry Plumb
240 Chas Bert Reed
240 G W Whitfield
241 Louise Dickerson
242 John B S King
271 John H DeWitt
362 William T Akins
362 M S Leech
367 M S Leech
518 Ebenezer Thurston
526 James H Stowell
1253 Albert H Burr
1332 J C Valentine
1332 Sarah L Valentine
1355 H W Joy
1412 Dayton Painter
1558 M G Hart
1700 J R Kewley
1734 Samnel Edwards

Taylor W.

144 F J Patera
397 William McCarthy
439 Patrick O'Connell
461 Geo C Synon
625 Gustav Schirmer
962 Ed Liebig
1153 Abbie J Mace

Thirteenth Pl.

529 E J Howe

Thirty-Eighth.

438 Charles E Caldwell

Thirty-Fifth.

52 James I Tucker
388 Albert L Clarke
390 E O Christoph
907 J H Krueger
1410 George Bell

Wabash ave.

1800 E Deane
1828 Ernest Schmidt
1828 O L Schmidt
2103 James E Gross
2103 M M Gross
2139 Rufus W Bishop
2139 Robert Randolph
2317 D Mills Tucker
2333 M H Landreth
2335 Louis N Barlow
2414 Chas Krusemarck
2515 Flavius M Wilder
2974 Emil Sincere
2979 Alfred Bosworth
3111 J R Kewley
3123 Martin Matter
3126 Harvey J Lyon
3155 Royce A Camp
3446 H Martin Scudder

Walnut.

38 Geo B Walker
50 W H Vary
161 A K Smith
262 J E Kennedy
658 W M W Davidson
853 A H Tagert

Warren ave.

9 Geo F Washburne
10 R N Foster
14 C A Williams
15 W M Stearns
90 Joseph Haven
91 L K Blakeslee
105 David W Graham
113 John M Fleming
191 Wells LeFevre
203 Odelia Blinn
259 V C McClure
259 Thomas D Palmer
330 H S Hann
730 Charles H Evans
747 Lucy E Ermine
821 Annette S Dobbin
856 Eli Wight

Washington.

85 Chas E Walker
85 J B Walker
103 George L Beach
108 John E Hurlbut
108 Frances E Thornton
108 Eli Wight
155 W W Hinish
163 C D Manning
170 E L Rivenburg

Washington W.

112 John M Fleming
300 Henrietta K Morris
306 Z P Hanson
315 W H Woodbury
325 Chas E Camp
325 Henry Van Buren
335 George W Reynolds
356 Henry C Feder
394 John Kean
428 J A Robinson
428 Joseph P Ross
435 Oscar C DeWolf
447 Watson Carr
455 John E Gilman
455 H N Lyon
471 John A McDonald
535 Charles Warrington Earle
535 Frank B Earle
582 Fred C Schaefer
636½ Philip Adolphus
652 W H Burt
681 John D Skeer
683 Albert E Hoadley
713 S J Avery
721 Sereno W Ingraham
726 C C Dodge
726 Walter F Knoll
751 E Garrott
811 A W Burnside
877 Gertrude Scott Condit

Washington Pl.

2 Junius M Hall
5 Roswell G Bogue
5 O N Huff

Washtenaw ave.

815 H A Coulter

Waverly Pl.

29 Marie J Mergler

Webster ave.

499 F H Foster

Wells.

82 O G Miller
94 Raymond L Leonard
138 N J Smedley
196 A Holmboe
226 C F P Korsell
248 C D B O'Ryan
322 A E Freund
324 Charles C Higgins
380 Theodore Strehz
397 H L Lemker
405 R Vampill
452 John Fisher
468 A Hottenroth
521 F A Karst

Wentworth ave.

2424 C Frithiof Larson
2427 Emil Schottenfels
2459 Alfred Dahlberg
2642 Edwin H Smith
2901 Charles Plaum
3100 A L Thomas
3109 J P Lynch
3901 Albert E Froom
4441 Otto Poppe

Western ave. S.

125 F S Horn
125 Philon C Whidden
143 J J M Angear
173 C Jackson Tobias
176 Jerome Henry Salisbury
255 C T Hood
358 A V Hutchins
1160 Geo W Boice

Western ave. N.

1061 J C Merrick

Willow.

83 A C Buell

Wilmot ave.

229 W W Hinish

Winchester ave.

122 W D Garvin

Winthrop Pl.

2 Donald Fraser

Wisconsin.

24 Charles H Berry
51-53 Ph H Mathei
65 Emilie Siegmund

Wood S.

114 Isaac Prince
114 Geo E Shipman
120 Geo E Shipman
193 A C Cotton
300 Albert J Ochener
300 H B Stehman
516 F McCormick

Wood N.

764 Geo Leininger.

STUDENT'S LIBRARY OF RUSH MED. COLLEGE

ILLINOIS MEDICAL PRACTICE ACT.

AN ACT TO REGULATE THE PRACTICE OF MEDICINE IN THE STATE OF ILLINOIS. APPROVED JUNE 16, 1887, IN FORCE JULY 12, 1887.

SECTION 1. *Be it enacted by the People of the State of Illinois, represented in the General Assembly,* That no person shall practice medicine in any of its departments in this State, unless such person possesses the qualifications required by this act. If a graduate in medicine, he shall present his diploma to the State Board of Health for verification as to its genuineness. If the diploma is found genuine, and from a legally chartered medical institution in good standing, and if the person named therein be the person claiming and presenting the same, the State Board of Health shall issue its certificate to that effect, signed by all the members thereof, and such certificate shall be conclusive as to the right of the lawful holder of the same to practice medicine in this State. If not a graduate, the person practicing medicine in this State shall present himself before said board and submit himself to such examination as the board may require, and if the examination be satisfactory to the board, the said board shall issue its certificate in accordance with the facts, and the lawful holder of such certificate shall be entitled to all the rights and privileges herein mentioned.

SEC. 2.* The State Board of Health shall organize within three months after the passage of this act, it shall procure a seal, and shall receive through its secretary applications for certificates and examinations; the president and secretary shall have authority to administer oaths, and the board to take testimony in all matters relating to its duties; it shall issue certificates to all who furnish satisfactory proof of having received diplomas or licenses from legally chartered medical institutions in good standing as may be determined by the board; it shall prepare three forms of certificates, one for persons in possession of such diplomas or licenses, the second for candidates examined and favorably passed on by the board, and a third for persons to whom certificates may be issued as hereinafter provided in section 12 of this act; it shall furnish to the county clerks of the several counties a list of all persons receiving certificates. In selecting places to hold its meetings, it shall, as far as is reasonable, accommodate applicants residing in differ nt sections of the State, and due notice shall be published of all of its meetings for examination. Certificates shall be signed by all the members of the board, and the secretary of the board shall receive from the applicant a fee of five (5) dollars for each certificate issued to such graduate or licentiate. Graduates or licentiates in midwifery to pay the sum of two (2) for each certificate. All such fees for certificates shall be paid by the secretary into the treasury of the board.

---

*See act establishing State Board of Health, *post.*

SEC. 3. The verification of the diploma shall consist in the affidavit of the holder and applicant that he is the lawful possessor of the same, and that he is the person therein named. Such affidavit may be taken before any person authorized to administer oaths, and the same shall be attested under the hand and official seal of such officer, if he have a seal; and any person swearing falsely shall be deemed guilty of perjury, and punished accordingly. Graduates may present their diplomas and affidavits as provided in this act, by letter or by proxy, and the State Board of Health shall issue its certificate the same as though the owner of the diploma was present.

SEC. 4. All examinations of persons not graduates or licentiates, shall be made directly by the board, and the certificates given by the board shall authorize the possessor to practice medicine and surgery in the State of Illinois.

SEC. 5. Every person holding a certificate from the State Board of Health shall have it recorded in the office of the clerk of the county in which he resides, within three months from its date, and the date of recording shall be indorsed thereon. Until such certificate is recorded as herein provided the holder thereof shall not exercise any of the rights or privileges conferred therein to practice medicine. Any person removing to another county to practice shall record the certificate in like manner in the county to which he removes, and the holder of the certificate shall pay to the county clerk the usual fees for making the record.

SEC. 6. The county clerk shall keep, in a book provided for the purpose, a complete list of the certificates recorded by him, with the date of the issue of the certificate. If the certificate be based on a diploma or license, he shall record the name of the medical institution conferring it, and the date when conferred. The register of the county clerk shall be open to public inspection during business hours.

SEC. 7. The fees for the examination of non-graduates shall be as follows: twenty (20) dollars for an examination in medicine and surgery; ten (10) dollars for an examination in midwifery only; and said fees shall be paid into the treasury of the board. If an applicant fails to pass said examination his or her fee shall be returned. Upon successfully passing the examination the certificate of the board shall be issued to the applicant without further charge.

SEC. 8. Examinations may be made in whole or in part writing, and shall be of an elementary and practical character, but sufficiently strict to test the qualifications of the candidate as a practitioner.

SEC. 9. The State Board of Health may refuse to issue the certificates, provided for in section 2 to individuals guilty of unprofessional or dishonorable conduct, and it may revoke such certificates for like causes. In all cases of refusal or revocation, the applicant may appeal to the Governor, who may affirm or overrule the decision of the board, and this decision shall be final.

SEC. 10. Any person shall be regarded as practicing medicine, within the meaning of this act, who shall treat, operate on, or prescribe for any physical ailment of another. But nothing in this act shall be construed to prohibit services in cases of emergency, or the domestic administration of family remedies. And this act shall not apply to commissioned surgeons of the United States Army, Navy or Marine Hospital Service in the discharge of their official duties.

SEC. 11. Any itinerate vendor of any drug, nostrum, ointment or appliance of any kind, intended for the treatment of disease or injury, or who shall, by writing or printing, or any other method, profess to cure or treat disease or deformity, by any drug, nostrum, manipulation or other expedient, shall pay a license of one hundred (100) dollars per month into the treasury of the board, to be collected by the State Board of Health, in the name of the People of the State of Illinois for the use of said Board of Health. And it shall be lawful for the State Board of Health to issue such license on application made to the State Board of Health, such license to be signed by the president of the board and attested by the secretary of the board, with the seal of the board. Any such itinerate vendor who shall vend or sell any such drug, nostrum, ointment or appliance without having a license so to do, shall, if found guilty, be fined in any sum not less than one hundred dollars, and not exceeding two hundred dollars for each offense, to be recovered in an action of debt before any court of competent jurisdiction. But such board may, for sufficient cause, refuse such license.

SEC. 12. * Any person practicing medicine or surgery in the State without the certificate issued by this board in compliance with the provisions of this act, shall for each and every instance of such practice forfeit and pay to the people of the State of Illinois, for the use of the said State Board of Health, the sum of one hundred (100) dollars for the first offense, and two hundred (200) dollars for each subsequent offense, the same to be recovered in an action of debt before any court of competent jurisdiction and any person filing, or attempting to file, as his own, the diploma or certificate of another, or a forged affidavit of identification, shall be guilty of a felony, and upon conviction, shall be subject to such fine and imprisonment as are made and provided by the statutes of the State for the crime of forgery: *Provided*, that all persons who have been practicing medicine continuously for ten years within this State prior to the taking effect of the act† to which this is an amendment, and who have not under said original act obtained a certificate from said Board of Health to practice medicine in this State, shall, upon proper application to said Board of Health receive such certificate, unless it shall be ascertained and determined by said Board of Health that the person so applying for a certificate is of immoral character, or guilty of unprofessional or dishonorable conduct, in which case, said Board of Health may reject such application: *And, provided*, that such application for a certificate shall be made within six months after the taking effect of this act, and all persons holding a certificate on account of ten years' practice shall be subject to all the requirements and discipline of this act, and the act to which this is an amendment, in regard to their future conduct in the practice of medicine the same as all other persons holding certificates, and all persons not having applied for or received such certificate within six months after the taking effect of this act, and all persons whose applications have for the causes herein named been rejected, or certificates revoked, shall, if they shall practice medicine, be deemed guilty of practicing in violation of law, and shall suffer the penalties herein provided.

SEC. 13. Upon conviction of either of the offenses mentioned in this act, the court shall, as part of the judgment, order that the defendant be committed to the common jail of the county until the fine and costs are paid, and upon failure to pay the same immediately, the defendant shall be committed under said order *Provided*, that either party may appeal in the same time and manner as appeals may be taken in other cases, except that where an appeal is prayed in behalf of the people, no appeal bond shall be required to be filed, whether the appeal be from a justice of the peace, or from the county or circuit court, or from the ap-

*Cases are now pending in the courts involving the constitutionality and construction of this section.

†Act approved May 29, 1877, and in force July 1, 1877. Section 17 of the act of 1877 provided "that the provisions of this act shall not apply to those that have been practing medicine ten years within this State."

pellate court. But it shall be sufficient, in behalf of the people of the State of Illinois, for the use of the State Board of Health, to pray an appeal, and thereupon appeal may be had without bond or security.

SEC. 14. All acts and parts of acts inconsistent, or in conflict with this act, are hereby repealed.

DEAD BODIES FOR DISSECTION.

An Act to Promote the Science of Medicine and Surgery in the State of Illinois.

SEC. 1. *Be it enacted by the People of the State of Illinois, represented in the General Assembly,* That superintendents of penitentiaries, houses of correction and bridewells, wardens of hospitals, insane asylums and poor houses, coroners, sheriffs, jailors, city and county undertakers, and all other State, county town and city officers in whose custody the body of any deceased person required to be buried at public expense shall be, shall give permission to any physician or surgeon (a licentiate of the State Board of Health) or to any medical college or school, public or private, of any city, town or county, upon his or their request therefor, to receive and remove free of charge or expense, after having given proper notice to relatives or guardians of the deceased, the bodies of such deceased persons to be buried at public expense, to be by him or them used within the State, for advancement of medical science, preference being given to medical colleges or schools, public or private; said bodies to be distributed to and among the same equitably: the number assigned to each being in proportion to the students of each college or school: *Provided, however,* that if any person claiming to be, and satisfying the proper authorities that he is of kindred of the deceased, shall ask to have the body for burial, it shall be surrendered for interment: *And, provided, further,* that any medical college or school, public or private, or any officers of the same, that shall receive the bodies of deceased persons for the purposes of scientific study, under the provisions of this act, shall furnish the same to students of medicine and surgery, who may be under their instruction, at a price not exceeding the sum of five dollars for each and every such deceased body so furnished.

SEC. 2. Any physician or surgeon (a licentiate of the Illinois State Board of Health) or any medical college or school, public or private, before receiving any dead body or bodies, shall give to the proper authority, surrendering the same to him or them, a sufficient bond that said body or bodies shall be used only for the promotion of medical science within this State; and whoever shall use said body or bodies for any other purpose or shall remove the same beyond the limits of this State, and whosoever shall sell or buy any such body or bodies or shall traffic in the same, shall be deemed guilty of a misdemeanor, and shall, on conviction, be fined in a sum of not less than one hundred dollars and be imprisoned in the county jail for a term of not less than thirty days nor more than one year; the fine accruing from such conviction, to be paid into the school fund of the county where the offense shall have been committed.

SEC. 3. Any officer refusing to deliver the remains or body of any deceased person when demanded in accordance with the provisions of this a t, shall pay a penalty of not less than fifty dollars nor more than one hundred dollars for the first offense, and for the second offense a penalty of not less than one hundred dollars, nor more than five hundred dollars, and for a third offense, or any offense thereafter, the penalty of not less than five hundred dollars, or be imprisoned in the county jail not less than six nor more than twelve months, or both, at the discretion of the court; such penalties to be sued for by the health department, as the case may be.

SEC. 4. It shall be the duty of preceptors, professors and teachers, and all officers of medical colleges or schools, public or private, who shall receive any dead body or bodies, in pursuance of the provisions of this act, decently to bury in some publc cemetery, or to cremate the same in a furnace properly constructed for that purpose, the remains of all bodies, after they shall have answered the purposes of study aforesaid, and for any neglect or violation of the provisions of this act, the party or parties so neglecting shall, on conviction, forfeit or pay a penalty of not less than fifty dollars, nor more than one hundred dollars, or be imprisoned in he county jail not less than six nor more than twelve months, or both, at the discretion of the court; such penalties to be sued for by school officers, or any person interested therein, for the benefit of the school fund of the county in which the offense shall have been committed.

SEC. 5. An act entitled "An act to promote the science of medicine and surgery in the State of Illinois," approved February 16, 1874, in force July 1, 1874, is hereby repealed.

Approved June 26, 1885. In force July 1st, 1885.

STATE BOARD OF HEALTH.

SECTION 1. *Be it enacted by the People of the State of Illinois, represented in the General Assembly*, that the Governor, with the advice and consent of the Senate, shall appoint seven persons who shall constitute the Board of Health. The persons so appointed shall hold their offices for seven years;

Provided, That the terms of office of the seven first appointed shall be so arranged that the term of one shall expire on the 30th day of December of each year; and the vacancies so created, as well as all vacancies occuring otherwise, shall be filled by the Governor, with the advice and consent of the Senate: and

Provided also, That appointments made when the Senate is not in session, may be confirmed at its next ensuing session.

SEC. 2. The State Board of Health shall have the general supervision of the interests of the health and life of the citizens of the State. They shall have charge of all matters pertaining to quarantine; and shall have authority to make such rules and regulations, and such sanitary investigations as they may, from time to time, deem necessary for the preservation or improvement of public health; and it shall be the duty of all Police Officers, Sheriffs, Constables, and all other officers and employes of the State, to enforce such rules and regulations so far as the efficiency and success of the Board may depend upon their official co-operation.

SEC. 3. The Board of Health shall have supervision of the State system of registration of births and deaths, as hereinafter provided; they shall make up such forms and recommend such legislation as shall be deemed necessary for the thorough registration of vital and mortnary statistics throughout the State. The Secretary of the Board shall be the Superintendent of such registration. The clerical duties, and the safe keeping of the bureau of vital statistics thus created shall be provided by the Secretary of State.

SEC 4. It shall be the duty of all Physicians and Accouchers in this State to register their names and postoffice address with the County Clerk of the County where they reside; and said Physicians and Accouchers shall be required, under penalty of $10 to be recovered in any court of competent jurisdiction in the

State, at suit of the County Clerk, to report to the County Clerk, within thirty days from the date of their occurrence, all births and deaths which may come under their supervision, with a certificate of the cause of death, and such correlative facts as the Board may require, in the blank forms as hereinafter provided.

SEC. 5. Where any birth or death shall take place, no Physician or Accoucher being in attendance, the same shall be reported to the County Clerk, within thirty days from the date of their occurrence, with the supposed cause of death, by the parent; or, if none, by the nearest of kin, not a minor; or, if none, by the resident householder where the death shall occur, under penalty as provided in the preceding section of this act.

SEC. 6. The Coroners of the several counties shall be required to report to the County Clerk all cases of death which may come under their supervision with the cause and mode of death, etc., as per forms furnished, under penalty as provided in section 4 of this act.

SEC. 7. All amounts recovered under the penalties herein provided shall be appropriated to a special fund for the carrying out the objects of this law.

SEC. 8. The County Clerks of the several Counties in the State shall be required to keep separate books for the registration of the names and post-office address of Physicians and Accouchers, for births, for marriages, and for deaths. Said books shall always be open to inspection without fee; and said County Clerks shall be required to render a full and complete report of all births, marriages and deaths to the Secretary of the Board of Health, annually, and at such other times as the Board may direct.

SEC. 9. It shall be the duty of the Board of Health to prepare such forms for the record of births, marriages and deaths as they may deem proper; the said forms to be furnished by the Secretary of said Board, to the County Clerks of the several Counties, whose duty it shall be to furnish them to such persons as are herein required to make reports.

SEC. 10. The first meeting of the Board shall be within fifteen days after their appointment, and thereafter in January and June of each year, and at such other times as the Board shall deem expedient. The meeting in January of each year shall be in Springfield. A majority shall constitute a quorum. They shall choose one of their number to be President, and they may adopt rules and by-laws for their government, subject to the provisions of this Act.

SEC. 11. They shall elect a Secretary, who shall perform the duties prescribed by the Board; and by this Act, he shall receive a salary which shall be fixed by the Board; he shall also receive his traveling and other expenses incurred in the performance of his official duties. The other members of the Board shall receive no compensation for their services but their traveling and other expenses, while employed on business of the Board, shall be paid. The President of the Board shall quarterly certify the amount due the Secretary, and on presentation of his certificate, the Auditor of State shall draw his warrant on the Treasurer for the amount.

SEC. 12. It shall be the duty of the Board of Health to make an Annual Report, through their Secretary, or otherwise, in writing, to the Governor of this State on or before the first day or January of each year, and such report shall include so much of the proceedings of the Board, and such information concerning vital statistics; such knowledge respecting diseases, and such instruction on the subject of Hygiene as may be thought useful by the Board, for dissemination among the people, with such suggestions as to legislative action as they may deem necessary.

Sec. 13. The sum of five thousand dollars ($5,000), or so much thereof as may be necessary, is hereby appropriated to pay the salary of the Secretary, meet the contingent expenses of the office of the Secretary, and the expenses of the Board, and all costs for printing, which together shall not exceed the sum hereby appropriated; said expenses shall be certified and paid in the same manner as that of the Secretary.

Sec. 14. The Secretary of State shall provide rooms suitable for the meetings of the Board, and office room for the Secretary.

In force July 1st, 1877.

ILLINOIS STATE BOARD OF HEALTH.

Members of the Board.—W. A. Haskell, M.D., President, Alton; W. R. Mackenzie, M.D., Chester; H. V. Ferrell, M.D., Carterville; Newton Bateman, LL. D., Galesburg; R. Ludlam, M.D., Chicago; A. L. Clark, M.D., Treasurer, Elgin; John H. Rauch, M.D., Secretary, Springfield.

INSTRUCTIONS.

Certificates authorizing the practice of Medicine and Surgery in the State of Illinois, are issued by the State Board of Health on complying with the following requirements, based upon the Amended Medical Practice Act, in force July 1, 1887:

1. The applicant must present to the State Board, at the office of the Secretary, in Springfield, or to any member of the Board, the *Diploma or License* of a legally chartered medical institution in *good standing*.—See *post* for definition of "Colleges in good standing." Such diploma or license may be presented in person, may be sent by mail or express. It is safer to send by express.

2. Said diploma or license must be accompanied by the *Affidavit* of the holder and applicant that he, or she, is the lawful possessor of the same, and is the person named therein.

3. No *Certificate* will be issued without *letters of recommendation* from reputable medical men with regard to the moral and professional character of the applicant—these to be filed as part of the record.

4. A fee of *five dollars* must be paid in advance. If not paid in person, a receipt should be obtained from the agency by which it is transmitted.

COLLEGES IN GOOD STANDING.

To be held in good standing in Illinois, Colleges must comply with the following schedule of requirements. This schedule took effect at the close of the session of 1882-83. No diploma issued since that date will be accepted as the basis for a Certificate, unless issued in accordance with this schedule.

The only other mode of obtaining a Certificate is through an examination by the Board.

SCHEDULE OF REQUIREMENTS.

I. Conditions of Admission to Lecture Courses.—1. Credible certificates of good moral standing. 2. Diplomas of graduation from a good literary and scientific college, or high school, or a first-grade teacher's certificate. Lacking such evidence of preliminary education—a thorough examination in the branches of a good English education, including mathematics, English composition, and elementary physics or natural philosophy.

II. Branches of Medical Science to be included in the Course of Instruction.—1. Anatomy. 2. Physiology. 3. Chemistry. 4. Materia Medica and Therapeutics. 5. Theory and Practice of Medicine. 6. Pathology. 7. Surgery. 8. Obstetrics and Gynæcology. 9. Hygiene. 10. Medical Judisprudence (Forensic Medicine).

III. Length of Regular or Graduating Courses.—1. The time occupied in the regular courses or sessions from which students are graduated shall not be less than five months, or twenty weeks, each. 2. Two full courses of lectures, not within one and the same year of time, shall be required for graduation with the Degree of Doctor of Medicine.

IV. Attendance and Examinations or Quizzes.—1. Regular attendance during the entire lecture courses shall be required, allowance being made only for absences occasioned by the student's sickness, such absence not to exceed twenty per centum of the course. 2. Regular examinations or quizzes to be made by each lecturer or professor daily, or at least twice each week. 3. Final examinations on all branches to be conducted, when practicable, by competent examiners other than the professors in each branch.

V. Dissections, Clinics and Hospital Attendance.—1. Each student shall have dissected during two courses. 2. Attendance during at least two terms of clinical and hospital instruction shall be required.

VI. Time of Professional Studies.—This shall not be less than three full years before graduation, including the time spent with a preceptor, and attendance upon lectures or at clinics and hospitals.

VII. Instruction.—The college must show that it has a sufficient and competent corps of instructors, and the necessary facilities tor teaching, dissections, clinics, etc.

INSTRUCTIONS.

FOR PRACTITIONERS OF TEN OR MORE YEARS IN THIS STATE, BEFORE JULY 1, 1887.*

Certificates authorizing the practice of Medicine and Surgery in the State of Illinois are issued by the State Board of Health, on complying with the following requirements, based upon the amended Medical Practice Act, in force July 1, 1887:

1. The applicant must file in the office of the Secretary of the Board, in Springfield, his affidavit that he *had practiced* Medicine and Surgery *continuously* in the State of Illinois *ten or more years* before July 1, 1877, and must also furnish corroborative evidence of the same.

2. No *Certificate* will be issued without *letters of recommendation* from reputable medical men, with regard to the moral and professional character of the applicant—these to be filed as part of the record.

3. A fee of *five dollars* must be paid in advance. If not paid in person, a receipt should be obtained from the agency through which it is transmitted.

*See Sec. 12 of Practice Act, limiting period within which application could be made to Jan. 1st, 1888.

MIDWIFE AFFIDAVIT.

INSTRUCTIONS.

If applicant has a certificate or diploma, it must be sent to the Board for verification, with letters of recommendation from reputable medical men; and if she has practiced in the State ten or more years before July 1, 1877, the affidavit must also be accompanied with letters of recommendation. Those without certificates or diplomas, and who have not practiced the required length of time in this State must be examined by the Board No certificates will be issued unless the requirements on the affidavit are complied with. The fee for examination is $5.00.

AN ACT.

TO INSURE THE BETTER EDUCATION OF PRACTITIONERS OF DENTAL SURGERY, AND TO REGULATE THE PRACTICE OF DENTISTRY IN THE STATE OF ILLINOIS. APPROVED MAY 30, 1881. IN FORCE JULY 1ST, 1881.

SECTION 1. *Be it enacted by the People of the State of Illinois, represented in the General Assembly:* That it shall be unlawful for any person, who is not at the time of the passage of this act engaged in the practice of dentistry in this State, to commence such practice, unless such person shall have received a diploma from the faculty of some reputable dental college, duly authorized by the laws of this State, or some other of the United States, or by the laws of some foreign country, in which college or colleges there was, at the time of the issue of such diploma, annually delivered a full course of lectures and instructions in dental surgery: *Provided*, that any person removing into this State, who shall have been, for a period of ten years prior to such removal, a practicing dentist, and provided, also, that any person holding the diploma of doctor of medicne from any reputable medical college, shall be entitled to practice dentistry in this State, upon obtaining a license for that purpose as hereinafter provided; and nothing in this act shall be construed to prohibit any physician or surgeon from extracting teeth.

SEC. 2. A board of examiners, to consist of five practicing dentists, is hereby created, whose duty it shall be to carry out the purposes and enforce the provisions of this act. The members of said board shall be appointed by the Governor. The term for which the members of said board shall hold their offices shall be five years, except that the members of the board first to be appointed under this act shall hold their offices for the term of one, two, three- four and five years, respectively, and until their successors shall be duly appointed. In case of a vacancy occurring in said board, such vacancy shall be filled by the Governor.

SEC. 3. Said board shall choose one of its members president, and one the secretary thereof, and it shall meet at least once in each year, and as much oftener and at such times and places as it may deem necessary. A majority of said board shall at all times constitute a quorum, and the proceedings thereof shall at all reasonable times be open to public inspection.

SEC. 4. It shall be the duty of every person who is engaged in the practice of dentistry in this State, within six months from the date of the passage of this act, to cause his or her name and residence or place of business to be registered with said board of examiners, who shall keep a book for that purpose; and every person who shall so register with said board as a practitioner of dentistry, may continue to practice the same as such, without incurring any of the liabilities or penalties provided in this act.

SEC. 5. No person whose name is not registered on the books of said board as a regular practitioner of dentistry, within the time prescribed in the preceding section, shall be permitted to practice dentistry in this State until such person shall have been duly examined by said board and regularly licensed in accordance with the provisions of this act.

SEC. 6. Any and all persons, who shall so desire, may appear before said board at any of its regular meetings and be examined with reference to their knowledge and skill in dental surgery, and if the examination of any such person or persons shall prove satisfactory to said board, the board of examiners shall issue to such persons as they shall find from such examination to possess the requisite qualifications, a license to practice dentistry, in accordance with the provisions of this act. But said board shall at all times issue a license to any regular graduate of any reputable dental college without examination, upon the payment by such graduate to the said board of a fee of one dollar. All licenses issued by said board shall be signed by the members thereof, and be attested by its president and secretary; and such license shall be *prima facie* evidence of the right of the holder to practice dentistry in the State of Illinois.

SEC. 7. Any member of said board may issue a temporary license to any applicant, upon the presentation by such applicant of the evidence of the necessary qualifications, to practice dentistry, and such temporary license shall remain in force until the next regular meeting of said board occurring after the date of such temporary license, and no longer.

SEC. 8. Any person who shall violate any of the provisions of this act shall be liable to prosecution, before any court of competent jurisdiction, upon information or by indictment, and upon conviction may be fined not less than twenty-five dollars, nor more than fifty dollars, for each and every offense. All fines recovered under this act shall be paid into the common school fund of the county in which such conviction takes place.

SEC. 9. In order to provide the means for carrying out and maintaining the provisions of this act, the said board of examiners may charge each person applying to or appearing before them for examination for license to practice dentistry, a fee of two dollars, and out of the funds coming into the possession of the board from the fees so charged, the members of said board may receive as compensation the sum of five dollars for each day actually engaged in the duties of their office and all legitimate and necessary expenses incurred in attending the meetings of said board. Said expenses shall be paid from the fees and penalties received by the board under the provisions of this act. And no part of the salary or other expenses of the board shall ever be paid out of the State treasury. All moneys received in excess of said per diem allowance, and other expenses above provided for, shall be held by the secretary of said board as a special fund for meeting the expenses of said board, he giving such bond as the board shall from time to time direct. And said board shall make an annual report of its proceedings to the Governor, by the fifteenth of December of each year, together with an account of all moneys received and disbursed by them pursuant to this act.

SEC. 10. Any person who shall be licensed by said board to practice dentistry, shall cause his or her license to be registered with the county clerk of any county or counties in which such person may desire to engage in the practice of dentistry, and the county clerks of the several counties in this State shall charge for registering such license a fee of twenty-five cents for each registration. Any failure, neglect or refusal on the part of any person holding such license to register the same with the county clerk, as above directed, for a period of six months, shall work a forfeiture of the license, and no license, when once forfeited, shall be restored, except upon the payment to the said board of examiners of the sum of twenty-five dollars, as a penalty for such neglect, failure or refusal.

ILLINOIS STATE BOARD OF DENTAL EXAMINERS.

Members.—R. N. Lawrence, D. D. S., Lincoln, President; G. V. Black, M. D., D. D. S., Jacksonville; Homer Judd, M. D., D. D. S., Upper Alton; C. A. Kitchell, Rockford; C. R. E. Koch, Chicago, Secretary. Office of Secretary, 3011 Indiana Ave.; hours, 8 a. m. to 1 p. m.

The above law, as construed by the Attorney General, makes it obligatory upon every dentist who commenced the practice of dentistry in the State of Illinois after July 1, 1881, or who shall hereafter commence such practice, to obtain a license from the board of examiners. Those dentists who hold diplomas from reputable* dental colleges can obtain license by proving their diplomas and paying a fee of one dollar for the same.

All others must appear before the board for examination.

All persons not holding diplomas, and desiring to commence practice before the regular meeting of the board of examiners, must apply to some member of the board for examination, for temporary license, which, if issued, will be valid until the next meeting of the board, at which time the person holding a temporary license must appear before the board for final examination.

The fee for each examination is two dollars. Examinations will be conducted in writing in Theory and Practice of Dentistry, and Clinical Dentistry; Anatomy, Histology and Surgery; Physiology, Pathology, Therapeutics and Materia Medica, Chemistry and Metallurgy, Prosthetic Dentistry, Deformities and Hygiene.

Practical demonstrations in operative dentistry will also be required.

Seventy-five per cent. of the whole number of questions must be correctly answered to entitle a person to a license; *but* any candidate who fails to answer correctly seventy-five per cent. of the questions in either of the following subjects, namely: *Theory and Practice, Anatomy, Physiology, Pathology and Surgery*, even though his general average should be seventy-five per cent. on the whole number of questions asked, will *not* be entitled to a license.

No persons who are not registered on the books of the board, or who do not hold license, can legally practice dentistry in this State, and if they attempt to do so, will be liable to prosecution for such illegal practice.

---

*The following resolution was adopted at the annual meeting held in Springfield, September, 1884.

Resolved, That after June, 1885, The Illinois State Board of Dental Examiners will recognize as reputable *only* such dental colleges as require as a requisite for graduation, attendance upon two full regular courses of lectures and practical instruction, which courses shall each be of not less than five months' duration, and shall be held in separate years, with practical instruction intervening between the courses. Such colleges must also require a preliminary examination before admitting students to matriculation; provided, that no certificate from a high or normal school or other literary institution, is presented by the candidate.

The regular meetings of the board will be held on the third Monday of September, in the State House, at Springfield, and on the second Monday of May, at such place as the Illinois State Dental Society may meet, due notice of which may be found in the Dental journals.

AN ACT.

TO REGULATE THE PRACTICE OF PHARMACY IN THE STATE OF ILLINOIS—APPROVED MAY 30, 1881—IN FORCE JULY 1, 1881.

Be it enacted by the People of the State of Illinois, represented in the General Assembly,

SECTION 1. That it shall not be lawful for any person other than a registered pharmacist to retail, compound or dispense drugs, medicines or poisons, or to open or conduct any pharmacy or store for retailing, compounding or dispensing drugs, medicines or poisons, unless such persons shall be, or shall employ and place in charge of said pharmacy or store a registered pharmacist, within the meaning of this act, except as hereinafter provided.

SEC. 2. Any person, in order to be registered within the meaning of this act, must be either a graduate in pharmacy, a graduate in medicine, or shall, at the time this act takes effect, be engaged in the business of a dispensing pharmacist on his own account, in the State of Illinois, in the preparation of physicians' prescriptions and in the vending and compounding of drugs, medicines and poisons, or shall be a licentiate in pharmacy.

SEC. 3. Graduates in pharmacy must be such persons as have had four years' practical experience in drug stores where the prescriptions of medical practitioners are compounded, and have obtained a satisfactory diploma or credentials of their attainments from a regular incorporated college or school of pharmacy.

SEC. 4. Licentiates in pharmacy must be such persons as have had two years' practical experience in drug stores where the prescriptions of medical practitioners are compounded, and have passed a satisfactory examination before the State Board of Pharmacy, hereinafter mentioned. The said Board may grant certificates of registration, without further examination, to the licentiates of such other Boards of Pharmacy as it may deem proper.

SEC. 5. The Governor, with the advice and consent of the Senate, shall appoint five persons from among such competent pharmacists in the State as have had ten years' practical experience in the dispensing of physicians' prescriptions who shall constitute the Board of Pharmacy. The persons so appointed shall hold their offices for five years: *Provided*, that the term of office of the five first appointed shall be so arranged that the term of one shall expire on the thirtieth day of December of each year; and the vacancies so created, as well as all vacancies otherwise occurring, shall be filled by the Governor, with the advice and consent of the Senate: *And, provided, also*, that appointments made when the Senate is not in session may be confirmed at its next ensuing session. The Illinois Pharmaceutical Association shall annually report directly to the Governor, recommending the first year the names of at least ten persons, whom said association shall deem best qualified to serve as members of the Board of Pharmacy, and the names of at least three persons each year thereafter, to fill any vacancies which shall occur in said Board.

SEC. 6. The said Board shall, within thirty days after its appointment, meet and organize by the election of a President and Secretary, from its own members, who shall be elected for the term of one year, and shall perform the duties prescribed by the Board. It shall be the duty of the Board to examine all applications for registration submitted in proper form; to grant certificates of registration to such persons as may be entitled to the same under the provisions of this act; to cause the prosecution of all persons violating its provisions; to report annually to the Governor and to the Illinois Pharmaceutical Association upon the condition of Pharmacy in the State, which said report shall also furnish a record of the proceedings of the said Board for the year, and also the names of all pharmacists duly registered under this act. The Board shall hold meetings for the examination of applicants for registration, and the transaction of such other business as shall pertain to its duties, at least once in three months: *Provided*, that said Board shall hold meetings once in every year in the city of Chicago and in the city of Springfield, and it shall give thirty days' public notice of the time and place of such meetings; shall have power to make by-laws for the proper fulfillment of its duties under this act, and shall keep a book of registration, in which shall be entered the names and places of business of all persons registered under this act, which book shall also specify such facts as said persons shall claim to justify their registration. Three members of said Board shall constitute a quorum.

SEC. 7. *Every person claiming the right of registration under this act who shall, within three months after this act shall take effect, forward to the Board of Pharmacy satisfactory proof, supported by his affidavit, that he was engaged in the business of a dispensing pharmacist on his own account in this State at the time this act takes effect, as provided in Section Two, shall, upon the payment of the fee hereinafter mentioned, be granted a certificate of registration: *Provided*, that in case of failure or neglect to register as herein provided, then such person shall, in order to be registered, comply with the requirements provided for registration as a graduate in pharmacy or a licentiate in pharmacy within the meaning of this act.

SEC. 8.—Any assistant or clerk in pharmacy, who shall not have the qualification of a registered pharmacist within the meaning of this act, not less than eighteen years of age, who, at the time this act takes effect, shall have been employed or engaged two years or more in drug stores where the prescriptions of medical practitioners are compounded, and shall furnish satisfactory evidence to that effect to the State Board of Pharmacy, shall, upon making application for registration, and upon payment to the Secretary of the said Board of a fee of one dollar, within sixty days after this act takes effect, be entitled to a certificate as a "registered assistant," which said certificate shall entitle him to continue in such duties as clerk or assistant: but such certificate shall not entitle him to engage in business on his own account unless he shall have had at least five years' experience in pharmacy at the time of the passage of this act. Annually thereafter, during the time he shall continue in such duties, he shall pay to the said Secretary a sum not exceeding fifty cents, for which he shall receive a renewal of his certificate.

SEC. 9. Every person applying for registration as a registered pharmacist, under Section Seven of this act, shall, before a certificate is granted, pay to the Secretary of the Board the sum of two dollars, and a like sum shall be paid to said Secretary by graduates in pharmacy, [by] graduates of medicine, and by licentiates of other boards who shall apply for registration; and by every ap-

*See amendment of this section, *post*.

plicant for registration by examination shall be paid the sum of five dollars: *Provided*, that in case of the failure of any applicant to pass a satisfactory examination his money shall be refunded.

SEC. 10. Every registered pharmacist who desires to continue the practice of his profession, shall annually thereafter, during the time he shall continue in such practice, on such date as the Board of Pharmacy may determine, pay to the Secretary of the said Board a registration fee, to be fixed by the Board, but which shall in no case exceed two dollars, for which he shall receive a renewal of said registration. Every certificate of registration granted under this act shall be conspicuously exposed in the pharmacy to which it applies.

SEC. 11. The Secretary of the Board shall receive a salary, which shall be fixed by the Board; he shall also receive his traveling and other expenses incurred in the performance of his official duties. The other members of the Board shall receive the sum of five dollars for each day actually engaged in this service, and all legitimate and necessary expenses incurred in attending the meetings of said Board. Said expenses shall be paid from the fees and penalties received by the Board, under the provisions of this act, and no part of the salary or other expenses of the Board shall be paid out of the State treasury. All moneys received in excess of said per diem allowance and other expenses above provided for, shall be held by the Secretary as a special fund for meeting the expenses of said Board, he giving such bonds as the Board shall from time to time direct. The Board shall, in its annual report to the Governor and to the Illinois Pharmaceutical Association, render an account of all moneys received and disbursed by them pursuant to this act.

SEC. 12. Any person not being, or having, in his employ a registered pharmacist, within the meaning of this act, who shall, sixty days after this act takes effect keep a pharmacy, or store for retailing or compounding medicine, or who shall take, use or exhibit the title of a registered pharmacist, shall, for each and every such offense, be liable to a penalty of fifty dollars. Any registered pharmacist who shall permit the compounding and dispensing of prescriptions, or the vending of drugs, medicines or poisons in his store or place of business, except under the supervision of a registered pharmacist, or except by a "registered assistant" pharmacist, or any pharmacist or "registered assistant," who, while continuing in business, shall fail or neglect to procure his annual registration, or any person who shall willfully make any false representation to procure registration for himself or any other person, shall, for every such offense, be liable to a penalty of fifty dollars: *Provided*, that nothing in this act shall apply to nor in any manner interfere with the business of any physician, or prevent him from supplying to his patients such articles as may seem to him proper, nor with the making or vending of patent or proprietary medicines, or medicines placed in sealed packages, with the name of the contents and of the pharmacist or physician by whom prepared or compounded, nor with the sale of the usual domestic remedies by retail dealers, nor with the exclusively wholesale business of any dealers except as hereinafter provided: *And provided, further*, that no part of this section shall be so construed as to give the right to any physician to furnish any intoxicating liquor as a beverage, on prescription or otherwise.

SEC. 13. No person shall add to or remove from any drug, medicine, chemical or pharmaceutical preparation, any ingredient or material for the purpose of adulteration or substitution, or which shall deteriorate the quality, commercial value or medicinal effect, or which shall alter the nature or composition of such drug, medicine, chemical or pharmaceutical preparation, so that it will not correspond to the recognized tests of identity or purity. Any person who shall thus wilfully adulterate or alter, or cause to be adulterated or altered, or shall sell or offer for sale any such adulterated or altered drug, medicine, chemical or pharmaceutical preparation, or any person who shall substitute, or cause to be substituted, one material for another, with the intention to defraud or deceive the purchaser, shall be guilty of misdemeanor, and be liable to prosecution under this act. If convicted, he shall be liable to all the costs of the action and all expenses incurred by the Board of Pharmacy in connection therewith, and for the first offense be liable to a fine of not less than fifty dollars nor more than one hundred dollars, and for each subsequent offense a fine of not less than seventy-five nor more than one hundred and fifty dollars. On complaint being entered, the Board of Pharmacy is hereby empowered to employ an analyst or chemist expert, whose duty it shall be to examine into the so claimed adulteration, substitution or alteration, and report upon the result of his investigation; and if said report justify such action, the Board shall duly cause the prosecution of the offender, as provided in this law.

SEC. 14. No person shall sell at retail any poison commonly recognized as such, and especially aconite, arsenic, belladonna, biniodide of mercury, carbolic acid, chloral hydrate, chloroform, conium, corrosive sublimate, creosote, croton oil, cyanide of potassium, digitalis, hydrocyanic acid, laudanum, morphine, nux vomica, oil of bitter almonds, opium, oxalic acid, strychnine, sugar of lead, sulphate of zinc, white precipitate, red precipitate, without affixing to the box, bottle, vessel or package containing the same, and to the wrapper or cover thereof, a label bearing the name of the article, and the word "poison" distinctly shown, with the name and place of business of the seller; he shall not deliver any of said poisons to any person under the age of fifteen years, nor shall he deliver any of said poisons to any person, without satisfying himself that such poison is to be used for a legitimate purpose: *Provided*, that nothing herein contained shall apply to the dispensing of physicians' prescriptions of any of the poisons or articles aforesaid. Any person failing to comply with the requirements of this section shall be liable to a penalty of five dollars for each and every such offense.

SEC. 15. All suits for the recovery of the several penalties prescribed in this act shall be prosecuted in the name of the "People of the State of Illinois," in any court having jurisdiction; it shall be the duty of the State's Attorney of the county where such offense is committed, to prosecute all persons violating the provisions of this act, upon proper complaint being made. All penalties collected under the provisions of this act shall inure, one-half to the Board of Pharmacy, and the remainder to the school fund of the county in which the suit was prosecuted and judgment obtained.

AN ACT TO AMEND SECTION SEVEN OF AN ACT ENTITLED "AN ACT TO REGULATE THE PRACTICE OF PHARMACY IN THE STATE OF ILLINOIS."

SEC. 1. *Be it enacted by the people of the State of Illinois, represented in the General Assembly,* That section seven (7) of an act entitled "An act to regulate the practice of pharmacy in the State of Illinois," approved May 30, 1881, in force July 1, 1881, be and the same is hereby amended so as to read as follows:

SEC. 7. Licentiates in pharmacy shall, at the time of passing their examination, be registered by the Secretary of the State Board of Pharmacy as registered pharmacists. Registered assistant pharmacists holding valid certificates as such may become registered as registered pharmacists upon making application to the Board of Pharmacy and paying a fee of two dollars therefor. No person shall hereafter be registered as a registered pharmacist except registered assistant pharmacists and registered pharmacists holding valid certificates as such, in force at the time this amendment takes effect, and licentiates in pharmacy.

(In force July 1, 1887.)

STATE BOARD OF PHARMACY.

Members of the Board.—Herman Schroeder, Quincy, term to expire December 30, 1888; Albert E. Ebert, 426 State St., Chicago, term to expire December 30, 1889; Chas. W. Day, Springfield, term to expire December 30, 1890; Francis A. Prickett, Carbondale, term to expire December 30, 1891; William P. Boyd, term to expire December 30, 1892. Herman Schroeder, President, Chas. W. Day, Secretary.

Regular meetings of the Board are held on the second Tuesday in January, April, July and October, alternately in Springfield and Chicago.

Special meetings are called as occasion may require.

The Board issues the following instructions: Registration is granted only on examination, and in order to be admitted to examination an applicant is required to furnish satisfactory proof that he is not less than eighteen years of age, and is either—

1st. A graduate of a College of Pharmacy in good standing; or

2d. A licentiate of another state having a Board of Pharmacy whose course of examination is approved by this Board, and that such applicant has had two years, experience in dispensing drugs at retail and filling prescriptions; or

3. That the applicant has had two years, experience in dispensing drugs and filling prescriptions, under the supervision of a registered pharmacist of this state, in a lawfully conducted retail drug store or pharmacy.

Blank applications may be obtained from the secretary and *must* be in his office at least five days prior to the examination, or they will *not be received.*

NATIONAL CODES OF ETHICS.

CODE OF ETHICS OF THE AMERICAN MEDICAL ASSOCIATION ADOPTED MAY, 1847.

OF THE DUTIES OF PHYSICIANS TO THEIR PATIENTS, AND OF THE OBLIGATIONS OF PATIENTS TO THEIR PHYSICIANS.

Article I. Duties of Physicians to their Patients.

SECTION 1. A physician should not only be ever ready to obey the calls of the sick, but his mind ought also to be imbued with the greatness of his mission, and the responsibility he habitually incurs in its discharge. Those obligations are the more deep and enduring, because there is no tribunal, other than his own conscience, to adjudge penalties for carelessness or neglect. Physicians should, therefore, minister to the sick with due impressions of the importance of their office; reflecting that the ease, the health and the lives of those committed to their charge depend on their skill, attention and fidelity. They should study, also, in their deportment, so as to unite *tenderness* with *firmness*, and *condescension* with *authority*, so as to inspire the minds of their patients with gratitude, respect and confidence.

SEC. 2. Every case committed to the charge of a physician should be treated with attention, steadiness and humanity. Reasonable indulgence should be granted to the mental imbecility and caprices of the sick. Secrecy and delicacy, when required by peculiar circumstances, should be strictly observed; and the familiar and confidential intercourse to which physicians are admitted in their professional visits should be used with discretion, and with the most scrupulous regard to fidelity and honor. The obligation of secrecy extends beyond the period of professional service; none of the privacies of personal and domestic life, no infirmity of disposition or flaw of character observed during professional attendance should ever be divulged by the physician, except when he is imperatively required to do so. The force and necessity of this obligation are indeed so great, that professional men have under certain circumstances, been protected in their observance of secrecy by courts of justice.

SEC. 3. Frequent visits to the sick are, in general, requisite, since they enable the physician to arrive at a more perfect knowledge of the disease—to meet promptly every change which may occur, and also tend to preserve the confidence of the patient. But unnecessary visits are to be avoided, as they give useless anxiety to the patient, tend to diminish the authority of the physician, and render him liable to be suspected of interested motives.

SEC. 4. A physician should not be forward to make gloomy prognostications, because they savor of empiricism, by magnifying the importance of his services in the treatment or cure of the disease. But he should not fail, on proper occasions, to give to the friends of the patient timely notice of danger, when it really occurs; and even to the patient himself, if absolutely necessary. This office, however, is so peculiarly alarming when executed by him, that it ought to be declined, whenever it can be assigned to any other person of sufficient judgment and delicacy. For the physician should be the minister of hope and comfort to the sick; that by such cordials to the drooping spirit he may smooth the bed of death, revive expiring life, and counteract the depressing influence of those maladies which often disturb the tranquility of the most resigned in their last moments. The life of a sick person can be shortened, not only by the acts, but also by the words or the manner of a physician. It is, therefore, a sacred duty to guard himself carefully in this respect, and to avoid all things which have a tendency to discourage the patient and depress the spirits.

SEC. 5. A physician ought not to abandon a patient because the case is deemed incurable; for his attendance may continue to be highly useful to the patient, and comforting to the relatives around him, even in the last period of a fatal malady, by alleviating pain and other symptoms, and by soothing mental anguish. To decline attendance, under such circumstances, would be sacrificing to fanciful delicacy and mistaken liberality that moral duty which is independent of and far superior to all pecuniary consideration.

SEC. 6. Consultations should be promoted in difficult or protracted cases, as they give rise to confidence, energy, and more enlarged views in practice.

SEC. 7. The opportunity which a physician not infrequently enjoys, of promoting and strengthening the good resolutions of his patients suffering under the consequences of vicious conduct, ought never to be neglected. His counsels, or even remonstrances, will give satisfaction, not offense, if they be proffered with politeness, and evince a genuine love of virtue, accompanied by a sincere interest in the welfare of the person to whom they are addressed.

Art. II. Obligations of Patients to Their Physicians.

SEC. 1. The members of the medical profession, upon whom is enjoined the performance of so many important and arduous duties toward the community, and who are required to make so many sacrifices of comfort, ease and health for the welfare of those who avail themselves of their services, certainly have a right to expect and require that their patients should entertain a just sense of the duties which they owe to their medical attendants.

SEC. 2. The first duty of a patient is to select as his medical adviser one who has received a regular professional education. In no trade or occupation do mankind rely on the skill of an untaught artist; and in medicine, confessedly the most difficult and intricate of the sciences, the world ought not to suppose that knowledge is intuitive.

SEC. 3. Patients should prefer a physician whose habits of life are regular, and who is not devoted to company, pleasure, or to any pursuit incompatible with his professional obligations. A patient should also confide the the care of himself and family, as much as possible, to one physician; for a medical man who has become acquainted with the peculiarities of constitution, habits and predispositions of those he attends, is more likely to be successful in his treatment than one who does not possess that knowledge.

A patient who has thus selected his physician should always apply for advice in what may appear to him trivial cases, for the most fatal results often supervene on the slightest accidents. It is of still more importance that he should apply for assistance in the forming stage of violent diseases; it is to a neglect of this precept that medicine owes much of the uncertainty and imperfection with which it has been reproached.

SEC. 4. Patients should faithfully and unreservedly communicate to their physician the supposed cause of their disease. This is the more important, as many diseases of a mental origin stimulate those depending on external causes, and yet are only to be cured by ministering to the mind diseased. A patient should never be afraid of thus making his physician his friend and adviser; he should always bear in mind that a medical man is under the strongest obligations of secrecy. Even the female sex should never allow feelings of shame or delicacy to prevent their disclosing the seat, symptoms and causes of complaints peculiar to them. However commendable a modest reserve may be in the common occurrences of life, its strict observance in medicine is often attended with most serious consequences, and a patient may sink under a painful and loathsome disease, which might have been readily prevented had timely intimation been given to the physician.

Sec. 5. A patient should never weary his physician with a tedious detail of events or matters not appertaining to his disease. Even as relates to his actual symptoms, he will convey much more real information by giving clear answers to interrogatories, than by the most minute account of his own framing. Neither should he obtrude upon his physician the details of his business, nor the history of his family concerns.

Sec. 6. The obedience of a patient to the prescriptions of his physician should be prompt and implicit. He should never permit his own crude opinions as to their fitness to influence his attention to them. A failure in one particular may render an otherwise judicious treatment dangerous, and even fatal. This remark is equally applicable to diet, drink and exercise. As patients become convalescent, they are very apt to suppose that the rules prescribed for them may be disregarded, and the consequence, but too often, is a relapse. Patients should never allow themselves to be persuaded to take any medicine whatever that may be recommended to them by the self-constituted doctors and doctresses who are so frequently met with, and who pretend to possess infallible remedies for the cure of every disease. However simple some of their prescriptions may appear to be, it often happens that they are productive of much mischief, and in all cases they are injurious, by contravening the plans of treatment adopted by the physician.

Sec. 7. A patient should, if possible, avoid even the *friendly visits of a physician* who is not attending him—and when he does receive them, he should never converse on the subject of his disease, as an observation may be made witho y intention of interference, which may destroy his confidence in the course is pursuing, and induce him to neglect the directions prescribed to him. A patient should never send for a consulting physician without the express consent of his own medical attendant. It is of great importance that physicians should act in concert; for although their modes of treatment may be attended with equal success when employed singly, yet conjointly they are very likely to be productive of disastrous results.

Sec. 8. When a patient wishes to dismiss his physician, justice and common courtesy require that he should declare his reasons for so doing.

Sec. 9. Patients should always, when practicable, send for their physician in the morning, before his usual hours for going out; for, by being early aware of the visits he has to pay during the day, the physician is able to apportion his time in such a manner as to prevent an interference of engagements. Patients should also avoid calling on their medical adviser unnecessarily during the hours devoted to meals or sleep. They should always be in readiness to receive the visits of their physician, as the detention of a few minutes is often a serious inconvenience to him.

Sec. 10. A patient should, after his recovery, entertain a just and enduring sense of the value of the services rendered him by his physician; for these are of such a character that no mere pecuniary acknowledgement can repay or cancel them.

OF THE DUTIES OF PHYSICIANS TO EACH OTHER AND TO THE PROFESSION AT LARGE.

Article I.—Duties for the Support of Professional Character.

SECTION 1.—Every individual, on entering the profession, as he becomes thereby entitled to all its privileges and immunities, incurs an obligation to exert his best abilities to maintain its dignity and honor, to exalt its standing, and to extend the bounds of its usefulness. He should, therefore, observe strictly such laws as are instituted for the government of its members; should avoid all contumelious and sarcastic remarks relative to the faculty as a body; and while, by unwearied diligence, he resorts to every honorable means of enriching the science, he should entertain a due respect for his seniors, who have, by their labors, brought it to the elevated condition in which he finds it.

SEC. 2. It is not in accord with the interest of the public or the honor of the profession that any physician or medical teacher should examine or sign diplomas or certificates of proficiency for, or otherwise be specially concerned with, the graduation of persons whom they have good reason to believe intend to support and practice any exclusive and irregular system of medicine. (Sec. 2 was adopted in Richmond, 1881.)

SEC. 3. There is no profession from the members of which greater purity of character, and a higher standard of moral excellence is required, than the medical; and to attain such eminence is a duty every physician owes alike to his profession and to his patients. It is due to the latter, as without it he cannot command their respect and confidence, and to both, because no scientific attainments can compensate for the want of correct moral principles. It is also incumbent upon the faculty to be temperate in all things, for the practice of physic requires the unremitting exercise of a clear and vigorous understanding; and on emergencies, for which no professional man should be unprepared, a steady hand, an acute eye, and an unclouded head, may be essential to the well-being, and even to the life, of a fellow-creature.

SEC. 4. It is derogatory to the dignity of the profession to resort to public advertisements, or private cards, or handbills, inviting the attention of individuals affected with particular diseases—publicly offering advice and medicine to the poor, gratis, or promising radical cures; or to publish cases and operations in the daily prints, or to suffer such publications to be made; to invite laymen to be present at operations; to boast of cures and remedies; to adduce certificates of skill and success; or to perform any other similar acts. These are the ordinary practices of empirics, and are highly reprehensible in a regular physician.

SEC. 5. Equally derogatory to professional character is it for a physician to hold a patent for any surgical instrument or medicine; or to dispense a secret *nostrum*, whether it be the composition or exclusive property of himself or of others. For, if such nostrum be of real efficacy, any concealment regarding it is inconsistent with beneficence and professional liberality; and if mystery alone give it value and importance, such craft implies either disgraceful ignorance or fraudulent avarice. It is also reprehensible for physicians to give certificates attesting the efficacy of patent or secret medicines, or in any way to promote the use of them.

Art. II.—Professional Services of Physicians to Each Other.

SEC. 1. All practitioners of medicine, their wives, and their children while under the parental care, are entitled to the gratuitous services of any one or more of the faculty residing near them, whose assistance may be required. A physician afflicted with disease is usually an incompetent judge of his own case; and the natural anxiety and solicitude which he experiences at the sickness of a wife, a child, or any one who, by the ties of consanguinity, is rendered peculiarly dear to him, tend to obscure his judgment, and produce timidity and irresolution in his practice. Under such circumstances, medical men are peculiarly dependent upon each other, and kind offices and professional aid should always be cheerfully and gratuitously afforded. Visits ought not, however, to be obtruded officiously, as such unasked civility may give rise to embarassment or interfere with that choice on which confidence depends. But if a distant member of the faculty, whose circumstances are affluent, request attendance and an honorarium be offered, it should not be declined, for no pecuniary obligation ought to be imposed which the party receiving it would wish not to incur.

Art. III.- Of the Duties of Physicians as Respects Vicarious Offices.

SEC. 1. The affairs of life, the pursuit of health, and the various accidents and contingencies to which a medical man is peculiarly exposed, sometimes require him temporarily to withdraw from his duties to his patients, and to request some of his professional brethren to officiate for him. Compliance with this request is an act of courtesy, which should always be performed with the utmost consideration for the interest and character of the family physician, and when exercised for a short period, all the pecuniary obligations for such service should be awarded to him. But if a member of the profession neglect his business in quest of pleasure and amusement, he cannot be considered as entitled to the advantages of the frequent and long-continued exercise of this fraternal courtesy, without awarding to the physician who officiates the fees arising from the discharge of his professional duties.

In obstetrical and important surgical cases, which give rise to unusual fatigue, anxiety and responsibility, it is just that the fees accruing therefrom should be awarded to the physician who officiates.

Art. IV. Of the Duties of Physicians in Regard to Consultation.

SEC. 1. A regular medical education furnishes the only presumptive evidence of professional abilities and acquirements, and ought to be the only acknowledged right of an individual to the exercise and honors of his profession. Nevertheless, as in consultation the good of the patient is the sole object in view, and this is often dependent on personal confidence, no intelligent regular practitioner, who has a license to practice from some Medical Board of known and acknowledged respectability recognized by this Association, and who is in good moral and professional standing in the place in which he resides, should be fastidiously excluded from fellowship, or his aid refused in consultation, when it is requested by the patient. But no one can be considered as a regular practitioner or a fit associate in consultation whose practice is based upon an exclusive dogma, to the rejection of the accumulated experience of the profession, and of the aids actually furnished by anatomy, physiology, pathology, and organic chemistry.

SEC. 2. In consultations, no rivalship or jealousy should be indulged; candor, probity, and all due respect should be exercised towards the physician having charge of the case.

SEC. 3. In consultations, the attending physician should be the first to propose the necessary questions to the sick; after which the consulting physician should have the opportunity to make such further inquiries of the patient as may be necessary to satisfy him of the true character of the case. Both physicians should then retire to a private place for deliberation, and the one first in attendance should communicate the directions agreed upon to the patient or his friends, as well as any opinions which it may be thought proper to express. But no statement or discussion of it should take place before the patient or his friends, except in the presence of all the faculty attending, and by their common consent, and no opinions or prognostications should be delivered which are not the result of previous deliberation and concurrence.

SEC. 4. In consultation, the physician in attendance should deliver his opinion first; and when there are several consulting, they should deliver their opinions in the order in which they have been called in. No decision, however, should restrain the attending physician from making such variations in the mode of treatment as any subsequent unexpected change in the character of the case may dema d. But such variation, and the reason for it, ought to be carefully detailed at the next meeting in consultation. The same privilege belongs also to the consulting physician, if he is sent for in an emergency, when the regular attendant is out of the way, and similar explanations must be made at the next consultation.

SEC. 5. The utmost punctuality should be observed in the visits of physicians when they are to hold consultation together, and this is generally practicable, for society has been considerate enough to allow the plea of a professional engagement to take precedence of all others, and to be an ample reason for the relinquishment of any present occupation. But, as professional engagements may sometimes interfere, and delay one of the parties, the physician who first arrives should wait for his associate a reasonable period, after which the consultation should be considered as postponed to a new appointment. If it be the attending physician who is present, he will, of course, see the patient and prescribe; but if he be the consulting one, he should retire, except in case of emergency, or when he has been called from a considerable distance, in which latter case he may examine the patient, and give his opinion, in writing and under seal, to be delivered to his associate.

SEC. 6. In consultations, theoretical discussions should be avoided, occasioning perplexity and loss of time. For there may be much diversity of opinion concerning speculative points, with perfect agreement in those modes of practice which are founded, not on hypothesis, but on experience and observation.

SEC. 7. All discussion in consultation should be held as secret and confidential. Neither by words nor manner should any of the parties to a consultation assert or insinuate that any part of the treatment pursued did not receive his assent. The responsibility must be equally divided between the medical attendants—they must equally share the credit of success, as well as the blame of failure.

SEC. 8. Should any irreconcilable diversity of opinion occur when several physicians are called upon to consult together, the opinion of the majority should be considered as decisive; but the members be equal on each side, then the decision should rest with the attending physician. It may, moreover, sometimes happen that two physicians cannot agree in their view of the nature of a case, and the treatment to be pursued. This is a circumstance much to be deplored, and should always be avoided, if possible, by mutual concessions, as far as they can be justified by conscientious regard for the dictates of judgment. But, in e event of its occurrence, a third physician should, if practicable, be called to act as umpire; and if circumstances prevent the adoption of this course, it must be left to the patient to select the physician in whom he is most willing to confide. But, as every physician relies on the rectitude of his judgment, he should, when left in the minority, politely and consistently retire from any further deliberation in the consultation, or participation in the management of the case.

SEC. 9. As circumstances sometimes occur to render a special consultation desirable when the continued attendance of two physicians might be objectionable to the patient, the member of the faculty whose assistance is required in such cases should sedulously guard against all future unsolicited attendance. As such consultations require an extraordinary portion both of time and attention, at least a double honorarium may be reasonably expected.

SEC. 10. A physician who is called upon to consult, should observe the most honorable and scrupulous regard for the character and standing of the practitioner in attendance; the practice of the latter, if necessary, should be justified, as far as it can be, consistently with a conscientious regard for truth, and no hint or insinuation should be thrown out which could impair the confidence reposed in him, or affect his reputation. The consulting physician should also carefully refrain from any of those extraordinary attentions or assiduities which are too often practiced by the dishonest for the base purpose of gaining applause, or ingratiating themselves into the favor of families and individuals.

Art. V.—Duties of Physicians in Cases of Interference.

SEC. 1. Medicine is a liberal profession, and those admitted into its ranks should found their expectations of practice upon the extent of their qualifications, not on intrigue or artifice.

SEC. 2. A physician, in his intercourse with a patient under the care of another practitioner, should observe the strictest caution and reserve. No meddling inquiries should be made—no disingenious hints given relative to the nature and treatment of his disorder; nor any course of conduct pursued that may directly or indirectly tend to diminish the trust reposed in the physician employed.

SEC. 3. The same circumspection and reserve should be observed when, from motives of business or friendship, a physician is prompted to visit an individual who is under the direction of another practitioner. Indeed, such visits should be avoided, except under peculiar circumstances: and, when they are made, no particular inquiries should be instituted relative to the nature of the disease, or the remedies employed, but the topics of conversation should be as foreign to the case as circumstances will admit.

SEC. 4. A physician ought not to take charge of or prescribe for a patient who has recently been under the care of another member of the faculty in the same illness, except in cases of sudden emergency, or in consultation with the physician previously in attendance, or when the latter has relinquished the case, or been regularly notified that his services are no longer desired. Under such cir cumstances, no unjust or illiberal insinuations should be thrown out in relation to the conduct or practice previously pursued, which should be justified as far as candor and regard for truth and probity will permit; for it often happens that patients become dissatisfied when they do not experience immediate relief, and, as many diseases are naturally protracted, the want of success in the first stage of treatment affords no evidence of a lack of professional knowledge and skill.

SEC. 5. When a physician is called to an urgent case, because the family attendant is not at hand, he ought, unless his assistance in consultation be desired, to resign the care of the patient to the latter immediately upon his arrival.

Sec. 6. It often happens in case of sudden illness, or of recent accidents or injuries, owing to the alarm and anxiety of friends, that a number of physicians are simultaneously sent for. Under these circumstances, courtesy should assign the patient to the first who arrives, who should select from those present any additional assistance that he may deem necessary. In all such cases, however, the practitioner who officiates should request the family physician, if there be one, to be called, and unless his further assistance be requested, should resign the case to the latter on his arrival.

Sec. 7. When a physician is called to the patient of another practitioner, in consequence of the sickness or absence of the latter, he ought, on the return or recovery of the regular attendant, and with the consent of the patient, to surrender the case.

[The expression, "patient of another practitioner," is understood to mean a patient who may have been under the charge of another practitioner at the time of the attack of sickness, or departure from home of the latter, or who may have called for his attendance during his absence or sickness, or in any other manner given it to be understood that he regarded the said physician as his regular medical attendant.]

Sec. 8. A physician when visiting a sick person in the country may be desired to see a neighboring patient, who is under the regular direction of another physician, in consequence of some sudden change or aggravation of symptoms. The conduct to be pursued on such an occasion is to give advice adapted to present circumstances; to interfere no further than is absolutely necessary with the general plan of treatment; to assume no further direction, unless it be expressly desired; and, in this last case, to request an immediate consultation with the practitioner previously employed.

Sec. 9. A wealthy physician should not give advice *gratis* to the affluent, because his doing so is an injury to his professional brethren. The office of a physician can never be supported as an exclusively beneficent one: and it is defrauding, to some degree, the common funds for its support, when fees are dispensed with which might justly be claimed.

Sec. 10. When a physician who has been engaged to attend a case of midwifery is absent, and another is sent for, if delivery is accomplished during the attendance of the latter, he is entitled to the fee, but should resign the patient to the practitioner first engaged.

Art. VI.—Of Difference Between Physicians.

Sec. 1. Diversity of opinion and opposition of interest may, in the medical as in other professions, sometimes occasion controversy, and even contention Whenever such cases unfortunately occur, and cannot be immediately terminated, they should be referred to the arbitration of a sufficient number of physicians, or a *court-medical.*

Sec. 2. As a peculiar reserve must be maintained by physicians towards the public, in regard to professional matters, and as there exist numerous points in medical ethics and etiquette through which the feelings of medical men may be painfully assailed in their intercourse with each other, and which cannot be understood or appreciated by general society, neither the subject matter of such differences, nor the adjudication of the arbitrators, should be made public, as publicity in a case of this nature may be personally injurious to the individuals concerned, and can hard y fail to bring discredit on the faculty.

Art. VII.—Of Pecuniary Acknowledgments.

Some general rules should be adopted by the faculty in every town or district, relative to *pecuniary acknowledgements* from their patients; and it should be deemed a point of honor to adhere to these rules with as much uniformity as varying circumstances will admit.

OF THE DUTIES OF THE PROFESSION TO THE PUBLIC, AND OF THE OBLIGATIONS OF THE PUBLIC TO THE PROFESSION.

Art. I.—Duties of the Profession to the Public.

SEC. 1. As good citizens it is the duty of physicians to be ever vigilant for the welfare of the community, and to bear their part in sustaining its institutions and burdens; they should also be ever ready to give counsel to the public in relation to matters especially appertaining to their profession, as on subjects of medical police, public hygiene, and legal medicine. It is their province to enlighten the public in regard to quarantine regulations- the location, arrangement and dietaries of hospitals, asylums, schools, prisons, and similar institutions,—in relation to the medical police of towns, as drainage, ventilation, etc.,—and in regard to measures for the prevention of epidemic and contagious disease; and when pestilence prevails it is their duty to face the danger, and to continue their labors for the alleviation of suffering, even at the jeopardy of their own lives.

SEC. 2. Medical men should also be always ready, when called on by the legally constituted authorities, to enlighten coroner's inquests and courts of justice on subjects strictly medical- such as involve questions relating to sanity, legitimacy, murder by poisons or other violent means, and in regard to various other subjects embraced in the science of medical jurisprudence. But in these cases, and especially where they are required to make *post-mortem* examination, it is just, in consequence of the time, labor and skill required, and the responsibility and risk they incur, that the public should award them a proper honorarium.

SEC. 3. There is no profession by the members of which eleemosynary services are more liberally dispensed than the medical: but justice requires that some limit should be placed to the performance of such good offices. Poverty, professional brotherhood, and certain of the public duties referred to in the first section of this article, should always be recognized as presenting valid claims for gaatuitous services; but neither institutions endowed by the public or by rich individuals, societies for mutual benefit, for the insurance of lives, or for analogous purposes, nor any profession or occupation, can be admitted to possess such privilege. Nor can it justly be expected of physicians to furnish certificates of inability to serve on juries, to perform military duty, or to testify to the state of health of persons wishing to insure their lives, to obtain pensions, or the like, without a pecuniary acknowledgment. But to individuals in indigent circumstances such professional services should always be cheerfully and freely accorded.

SEC. 4. It is the duty of physicians, who are frequently witnesses of the enormities committed by quackery, and the injury to health, and even destruction of life, caused by the use of quack medicines, to enlighten the public on these subjects, to expose the injuries sustained by the unwary from the devices and pretensions of artful empirics and imposters. Physicians ought to use all the influence which they may possess, as professors in colleges of pharmacy, and by exercising their option in regard to the shops to which their prescriptions shall be sent, to discourage druggists and apothecaries from vending quack or secret medicine, or from being in any way engaged in their manufacture or sale.

Art. II.—Obligations of the Public to Physicians.

SECTION 1. The benefits accruing to the public, directly or indirectly, from the active and unwearied beneficence of the profession are so numerous and important that physicians are justly entitled to the utmost consideration and respect from the community. The public ought likewise to entertain a just appreciation of medical qualifications; to make a proper discrimination between true science and the assumption of ignorance and empiricism—to afford every encouragement and facility for the acquisition of medical education—and no longer to allow the statute-books to exhibit the anomaly of exacting knowledge from physicians under a liability to heavy penalties, and of making them obnoxious to punishment for resorting to the only means of obtaining it.

CODE OF ETHICS OF THE AMERICAN INSTITUTE OF HOMŒOPATHY.

SCOPE.

The scope of a Code of Medical Ethics comprises the reciprocal duties and obligations of physicians and patients; the duties and obligations of physicians to each other, and the reciprocal duties and obligations of physicians and the public.

FUNDAMENTAL PRINCIPLES.

The great principles upon which Medical Ethics are based are these:

1. The great end and object of the physician's efforts should be: "The greatest good to the patient."

2. The rule of conduct of physicians and patient, and of physicians towards each other, should be the GOLDEN RULE: "As ye would that men should do to you, do ye also to them likewise."

The various articles of the code are only special applications of these great principles.

PART I.

Reciprocal Duties and Obligations of Physicians and Patients.

ARTICLE I. Duties of Physicians to Patients.
ARTICLE II. Duties of Patients to Physicians.

PART II.

Duties and Obligations of Physicians to each other.

ARTICLE I. Duties as Members of the Medical Profession.
ART. II. Professional Services to Each Other.
ART. III. Vicarious Services.
ART. IV. In regard to Consultations.
ART. V. In cases of Interference.
ART. VI. Differences between Physicians.
ART. VII. Concerning Pecuniary Obligations.

PART III.

Reciprocal Duties and Obligations of Physicians and the Public.

ARTICLE I. Duties of the Profession to the Public.
ARTICLE II. Obligations of the Public to Physicians.

PART I.

OF THE RECIPROCAL DUTIES AND OBLIGATIONS OF PHYSICIANS AND THEIR PATIENTS.

Article I. Duties of the Physician to the Patient.

SECTION 1. The physician should hold himself in constant readiness to obey the calls of the sick. He should ever bear in mind the sacred character of his calling and the great responsibility which it involves, and should remember that the comfort, the health and the lives of his patients depend upon the skill, attention and faithfulness with which he performs his professional duties.

SEC. 2. The physician, in order that he may be able to exercise his vocation to the best advantage of the patient, should possess his respect and confidence. These must be acquired and retained by faithful attention to his malady, by indulgent tenderness towards the weaknesses incident to his condition, and by the exercise of a firm but kindly authority. The physician is bound to keep secret whatever he may either hear or observe, while in discharge of his professional duties, respecting the private affairs of the patient or his family. And this obligation is not limited to the period during which the physician is in attendance on the patient. The patient should be made to feel that he has, in the physician, a friend who will guard his secrets with scrupulous honor and fidelity.

SEC. 3. The physician should visit his patient as often as may be necessary to enable him to acquire and keep a full knowledge of the nature, progress, changes and complications of the disease, and to do for the patient the utmost of good that he is able. But he should carefully avoid making unnecessary visits, lest he render the patient needlessly anxious about his case, or expose himself to the charge of being actuated by mercenary motives.

SEC. 4. The physician should not give expression to gloomy forebodings respecting the patient's disease, nor magnify the gravity of the case. Bearing in mind the almost infinite resources of nature, he should be cheerful and hopeful, both in mind and manner. This will enable him the better to exercise his faculties and apply his knowledge for the patient's benefit, and will inspire the patient with confidence, courage and fortitude, which are the physician's best moral adjuvants.

But it is the physician's duty to state the true nature and prospects of the case, from time to time, to some judicious friend or relative of the patient, and to keep this person fully informed of its changes and probable issue; and if the patient himself request the physician to disclose to him the nature and prognosis of his disease, it is his duty to state, tenderly but frankly, the whole truth—provided the patient be of sound mind, and strong enough to receive the disclosure without serious injury. The patient has a right to know the truth. If, moreover, facts within the physician's knowledge lead him to believe that it is of great importance, in relation to the patient's affairs, that he should be warned of the approach of death, it is the physician's duty to reveal to the patient's nearest friend, or to the patient himself, the true state of the case, and the importance of timely action.

SEC. 5. Whether the case proceed favorably, or become manifestly incurable, it is the physician's duty to continue his attendance faithfully and conscientiously so long as the patient may desire it. He is not justified in abandoning a case merely because he supposes it incurable.

SEC. 6. As the patient has an undoubted right to dismiss his physician for reasons satisfactory to himself, so, likewise, the physician may, with equal propriety, decline to attend patients, when his self-respect or dignity seem to him to require this step; as, for example, when they persistently refuse to comply with his directions.

SEC. 7. In difficult or protracted cases, consultations are advisable. They tend to increase the knowledge, energy and confidence of the physician, and to maintain the courage of the patient. The physician should be ready to act upon any desire which the patient may express for a consultation, even though he may not himself feel the need of it. Nothing is so likely to maintain the patient's confidence as alacrity in this respect. Moreover, such a course is but just to him, for he has an indisputable right to whatever aid or counsel he may think likely to be of service to him.

SEC. 8. The intimate relations into which the physician is brought with his patient give him opportunity to exercise a powerful moral influence over him. This should always be exerted to turn him from dangerous or vicious courses towards a temperate and virtuous life. The physician is sometimes called to assist in practices of questionable propriety, and even a criminal character. Among these may be mentioned the pretense of disease, in order to evade services demanded by law, as jury or military duty; the concealment of organic disease or of morbid tendencies, in order to secure favorable rates of life insurance, or for deception of other kinds; and especially the procurement of abortion when not necessary to save the life of the mother. To all such propositions the physician should present an inflexible opposition. It is his duty, in an authoritative but friendly manner, to explain and urge the nature, illegality and guilt of the proposed action, and to use every effort to dissuade from it, and to strengthen the patient's virtue and sense of right. The physician should be aware of the frequency of criminal abortion, and of the different methods employed for it, and should take every occasion to warn those who may be tempted to resort to it. In no case should the physician induce abortion, or premature labor, without a previous consultation with the most experienced practitioners attainable, nor without the most clear and imperative reasons.

Art. II.—Duties and Obligations of Patients to their Physician.

SECTION 1. Physicians are required, by the nature of their profession, to sacrifice comfort, ease, and even health, for the sake of their patients. Patients should reflect upon this, and should understand and remember that they have corresponding duties and obligations towards their physicians.

SEC. 2. The patient should select a physician in whose knowledge, skill, and fidelity he can place implicit confidence; whose habits of life are regular and temperate, and whose character and demeanor are such that he can regard him as a personal friend. He must be able to confide in him freely. And the physician should not be changed for light reasons. A physician thoroughly acquainted with the constitutions, temperaments and tendencies of a family can the more successfully treat them.

SEC. 3. The patient should always consult his physician as early as possible after he has discovered that he is ill. A disease which is trifling at its onset may grow formidable through neglect. The physician should be regarded as a confidential adviser, who, on being early consulted, may prevent a sickness.

SEC. 4. The patient should faithfully and unreservedly state to his physician the supposed cause of his malady, and tell him everything that may have a bearing upon its nature. Since the physician is under the strongest obligations to secrecy, the patient should not allow considerations of delicacy, modesty, or pride to prevent an entirely frank statement of his case, and candid and full replies to interrogatories.

Sec. 5. The patient should implicitly obey his physician's injunctions as regards diet, regimen and medical treatment. If he deviates from these directions, he cannot hold the physician to a full responsibility in the case; and, further, by a partial obedience he incurs some personal risk, since, in the treatment of diseases, all parts of the physician's advice are made to harmonize, and each is dependent on the others and may be unsafe without the coincidence of the others. Moreover, he does the physician an undeserved and often a serious wrong. If the patient have not sufficient confidence in his physician, and respect for him, to follow his directions, it were better for him frankly to say so, and to employ another in whom he can confide.

The patient should never allow himself, while under a physician's treatment, to take other medicines than those prescribed by him. He would, by so doing, incur a serious risk of taking medicines that are incompatible with each other. If desirous of trying any other mode of treatment, it would be much better frankly to state the fact to his physician, and ask his advice.

Sec. 6. The patient should, if possible, avoid receiving the friendly visits of a physician other than the one under whose charge he is. When he receives such visits, he should avoid conversation on the subject of his disease; for an accidental observation might give him false impressions respecting his disease, or destroy his confidence in the treatment he is pursuing. He should never send for a consulting physician without the express consent of his own medical attendant; for physicians can act together for the advantage of their patient only when they act harmoniously. Nor should he, by a secret appointment, constrain his medical attendant to meet another physician with whom he might not be willing to consult; but the patient has an undoubted right to have the opinion of any physician whom he may desire upon his case. His proper course is to request his medical attendant to arrange a consultation, and frankly state his desire for the physician whom he may prefer. If his medical attendant decline the consultation, it is then for the patient to determine whether he will insist, and thus dismiss his medical attendant, or whether he will defer to the judgment of his own physician. And a patient has a right thus to choose.

Sec. 7. If the patient wishes to dismiss his physician, he should, in justice and common courtesy, state his reasons, and, if possible, in a friendly manner. To dispense with the services of a physician need not, of necessity, change the social relations of the parties.

Sec. 8. The patient should, when practicable, send for the physician in the morning, before his usual hour for leaving home. He will, by so doing, secure his earlier attendance, and will enable him the better to apportion his time so as to do justice to all his calls and engagements. He should call on his physician during his office hours only, and should avoid disturbing him in hours devoted to meals, rest and sleep. And in receiving his physician's visits, he should avoid compelling him to wait, even a few minutes. The aggregate of petty detentions, while the patient is making some needless preparation to receive the physician, amounts to a serious waste of valuable time.

PART II.

OF THE DUTIES AND OBLIGATIONS OF PHYSICIANS TO THE PROFESSION AND TO EACH OTHER.

Article I.—Duties to the Profession.

SECTION 1. Inasmuch as every member of the medical profession partakes of the honor in which it is held, is entitled to its privileges and immunities, and profits by the scientific labors of his predecessors and associates, it is his duty faithfully to endeavor, in his turn, to elevate the position of the profession and, by every honorable exertion, to enrich the science of medicine.

SEC. 2. In no other profession should a higher standard of morality and greater purity of personal character be required. Physicians ought to come up to this standard, and do what they may to exalt it. As the practice of medicine requires the constant exercise of a vigorous and clear understanding, and as the practitioner should be, at all times, ready for emergencies in which the welfare and even the life of a fellow creature may depend upon his steady hand, acute eye, and unclouded brain, it is incumbent upon the physician to be temperate in all things.

SEC. 3. The physician should not resort to public advertisements or private cards or handbills, inviting the attention of persons affected by particular diseases, or publicly offering advice and medicine to the poor, *gratis*, or promising radical cures. Neither should he publish cases or operations in the daily prints; nor invite laymen to be present at operations; nor solicit or exhibit certificates of skill and success; nor perform any similar act.

SEC. 4. It is equally derogatory to professional character for a physician to hold a patent for any nostrum or any surgical instrument or appliance; or to keep secret the nature and composition of any medicine used by him. Such restriction or concealment is inconsistent with the beneficence and liberality which should characterize the medical profession. But it is the duty of the physician to avail himself of every opportunity to observe the action and study the properties of new or secret remedies and new processes of preparing medicines as well as new modes of treating diseases, and to subject them to the analysis of scientific investigation. For the physician should always bear in mind that the great object of his profession is to cure the sick, and that it is not only admissible, but is his solemn duty to investigate, thoroughly and without prejudice, whatever offers any probability of adding to his knowledge of the art and means of curing, and of thus enriching the science of medicine.

Article II.—Professional Services of Physicians to each other.

SECTION 1. All practitioners of medicine, their wives, and children while under the paternal care, are entitled to the gratuitous services of any one or more of the faculty residing near them. Physicians, when ill, are incompetent to prescribe for themselves. The natural anxiety and solicitude which they feel for members of their own family when ill, tend to obscure their professional judgment and make it difficult to treat them. Under these circumstances, physicians are peculiarly dependent on each other; and kind offices and professional aid should always be cheerfully and gratuitously afforded. But visits should not be obtruded, officiously or unasked, upon a sick physician.

If, however, a physician in affluent circumstances request the attendance of a distant professional brother and offer an *honorarium*, it is not proper to decline it; for one should not, even from a kindly motive, impose upon another a pecuniary obligation which the recipient would not wish to incur.

If a physician is called from any considerable distance, the expense of travel, etc., thereby incurred, should always be paid by the physician receiving the visit; and an *honorarium* may be tendered if much time is consumed in making the visit.

Article III.—Duties of Physicians as Regards Vicarious Offices.

SECTION 1. Attention to his personal affairs, the pursuit of health, and the various contingencies to which the physician is peculiarly exposed, sometimes compel him temporarily to withdraw from his duties to his patients, and to request some of his professional brethren to discharge them for him. Compliance with such a request is an act of courtesy which should always be performed with the utmost consideration for the interests and character of the physician relieved. And when this is done for a short period only, all the pecuniary obligations for such services should belong to him. But if a physician neglect his business in quest of amusement and pleasure, he is not entitled to the frequent and long-continued exercise of this fraternal courtesy without conceding to the physician who acts for him the fees accruing from the duties discharged by the latter.

SEC. 2. Obstetrical and surgical cases involve unusual fatigue and responsibility; and it is just that the fees accruing therefrom should belong to the physician who attends them.

Article IV.—Duties of Physicians in Regard to Consultations.

SECTION 1. A complete medical education, of which the diploma of a medical college is the formal voucher, furnishes the only presumptive evidence of professional acquirements and abilities. But the annals of the profession contain the names of some who, not having the advantages of a complete medical education, became, nevertheless, through their own exertions and abilities, brilliant scholars and successful practitioners. A practitioner, therefore, whatever his credentials may be, who enjoys a good moral and professional standing in the community, should not be excluded from fellowship, nor his aid rejected, when it is desired by the patient in consultation. No difference in views on subjects of medical principles or practice should be allowed to influence a physician against consenting to a consultation with a fellow practitioner. The very object of a consultation is to bring together those who may, perhaps, differ in their views of the disease and its appropriate treatment, in the hope that, from a comparison of different views, may be derived a just estimate of the disease and a successful course of treatment.

No tests of orthodoxy in medical practice should be applied to limit the freedom of consultations. Medicine is a progressive science. Its history shows that what is heresy in one century may and probably will be orthodoxy in the next. No greater misfortune can befall the medical profession than the action of an influential association or academy establishing a creed or standard of orthodoxy or "regularity." It will be fatal to freedom and progress in opinion and practice. On the other hand, nothing will so stimulate the healthy growth of the profession, both in scientific strength and in the honourable estimation of the public, as the universal and sincere adoption of a platform which shall recognize and guarantee:—

1. A truly fraternal good-will and fellowship among all who devote themselves to the care of the sick.

2. A thorough and complete knowledge, however obtained, of all the direct

and collateral branches of medical science, as it exists in all sects and schools of medicine,—as the essential qualification of a physician.

3. Perfect freedom of opinion and practice, is the unquestionable prerogative of the practitioner, who is the sole judge of what is the best mode of treatment in each case of sickness intrusted to his care.

The physician may, with propriety, decline to meet a practitioner of whose inimical feelings towards himself or of whose general unfairness in consultations he is satisfied. But, in such a case, he should explain to the patient his reasons; and if the patient desire the opinion of the practitioner objected to, the family physician may withdraw from the case and allow the other to be sent for. But, in justice to the latter, the state of affairs should be explained to him at the time he is requested to visit the patient.

SEC. 2. The utmost punctuality should be observed in the visits of physicians when they are to hold consultations together; and this is generally practicable, for society allows the plea of professional engagements to excuse the neglect of all others, and to be a valid reason for the relinquishment of any present occupation. But, as professional engagements may sometimes interfere and delay one of the parties, the physician who first arrives should wait for his associate a reasonable period of time, after which the consultation should be considered postponed to a new appointment. If it be the attending physician who is present, he will, of course, see the patient and prescribe; but if it be the consulting physician he should retire without seeing the patient, except in cases of emergency or when he has been called from a considerable distance, in which case he may examine the patient, and give his opinion in writing and under seal, to be delivered to the attending physician.

SEC. 3. In consultations, no rivalry or jealously should be indulged in. Candor, probity, and all due respect should be exercised toward the physician in charge of the case. If the consulting physician cannot agree with him respecting the nature and proper treatment of the case, the physician should state this fact to the patient, or his nearest friend, both physicians being present at the time, and should request him to select the one in whom he has the most confidence. But, if they agree sufficiently to take joint charge of the case, then the consulting physician must justify and uphold, so far as he can conscientiously do so, the practice of his associate, and must abstain from any hints, insinuations or actions which might, in any way, impair the confidence which the patient reposes in him, or affect his reputation. He must refrain from any extraordinary attentions or assiduities, calculated to ingratiate himself in the patient's favor and to supplant his associate.

SEC. 4. In consultations, the attending physician should first put the necessary questions to the patient. After this, the consulting physician should make such additional inquiries and examinations as may be needed to satisfy him of the true nature of the case. But he should avoid making a parade of examining the patient more thoroughly than had been done before; rather suggesting to the attending physician, where this is possible, to make whatever examinations he desires, than making them himself. Both physicians should then retire to a private room for deliberation.

SEC. 5. In consultation the attending physician should deliver his opinion first; and, when there several consulting physicians, they should express their opinions in the order in which they have been called in. Should an irreconcilable diversity of opinion occur, when more than two physicians meet in consultation, the opinion of the majority should be regarded as decisive; but, if the number be equal on each side, the decision should rest with the attending physician. If two physicians, in consultation, cannot agree, they should call in a third to act as umpire. If this be not practicable, the patient must be requested to select the physician in whom he is most willing to confide. The physician who is left in the minority should, without any ill-feeling, retire from the consultation and from any further participation in the management of the case; and, in justice to the physician thus retiring, the fact of his difference from his associates should, in the presence of all the physicians attending, be explained to the patient, as his reason for withdrawing from the case.

SEC. 6. The attending physician should communicate to the patient or his friends the directions agreed upon in the consultation, as well as any opinion which it may be thought proper to express. But no statement or discussion should take place before the patient or his friends, except in the presence of all the physicians attending, and by their common consent. And no opinions or prognostications should be delivered, which are not the result of previous deliberation and concurrence. No decision arrived at in a consultation is to be regarded as restraining the attending physician from making such variations in the treatment as any subsequent change in the case may demand. But such variation and the reasons for it, ought to be carefully noted at the time, and detailed at the next meeting in consultation. The same privilege belongs also to the consulting physician, if he is sent for in an emergency when the attending physician is out of the way; and similar explanations must be made by him at the next meeting.

SEC. 7. Sometimes a special consultation is desirable in cases in which the continued attendance of two physicians might be objectionable to the patient. The consulting physician, in such a case, should sedulously avoid all further unsolicited attendance. Such consultations require an extraordinary outlay of time and attention, and at least a *double honorarium* may be reasonably expected.

SEC. 8. The consulting physician cannot, with propriety, take exclusive charge, at any time, of the patient in whose case he has been called in consultation, without the consent of the attending physician, except in cases provided for by the third sentence of section 3, and the fourth sentence of section 5, of this article.

Article V.—Duties of Physicians in Cases of Interference.

SEC. 1. Medicine is a liberal profession, and those admitted into its ranks should base their expectations of success upon the extent of their qualifications, not upon intrigue or artifice. A physician should not allow himself to feel envious or jealous of a brother-practitioner. The distinction which one successful physician wins is shared by the whole profession. Nor should a physician suffer himself to feel ill-will towards another who may come into his neighborhood and appears likely to take a share of the business which he has hitherto enjoyed. Such feelings are inconsistent with the beneficent and liberal nature of the profession. Liberality, and true generous fraternity in thought, word and deed, will unite the interests of all the members of the profession, and will so exalt the estimation in which it is held in the community that, confidence being increased, business will likewise increase; and to physicians will be accorded the position which, of right, should be theirs; that of confidential family advisers in all matters pertaining to the care of the body in health, no less than in sickness.

SEC. 2. The physician, in his intercourse with a patient who is under the care of another practitioner, should observe the strictest caution and reserve.

No meddling questions should be asked in any interview for business or friendship, no disingenuous hints thrown out relating to the nature and treatment of his disorder; nor should the patient be allowed to converse upon these topics. No course of conduct should be pursued which might, directly or indirectly, tend to diminish the trust reposed in the physician employed.

SEC. 3. A physician should not take charge of a patient who is, or has recently been, under the charge of another practitioner in the same illness, except in cases of sudden emergency, or in consultation with the physician previously in attendance, or when the latter has relinquished the case, or has been regularly notified that his services are no longer required. Under such circumstances no unjust or illiberal remarks should be made or insinuations thrown out in relation to the treatment pursued by the previous physician. Nor should the physician permit the patient unreasonably to find fault with his predecessor. For patients often become dissatisfied with their attendant on account of the mere duration of a case which no degree of professional knowledge or skill could have shortened.

SEC. 4. In cases of accident or sudden emergency, one or more physicians are often sent for by alarmed friends. Courtesy should assign the patient to the first of these that arrives; and he should select from those present such additional assistance as he may deem necessary. But he should also request the family physician (if there be one) to be sent for, and, on his arrival, resign the case into his hands. The practitioner of the patient, when he arrives, should take the place of any one called in his absence. "The practitioner of any patient" is the man whom he has in any way given to understand that he regards him as his medical adviser, or who would now be in charge of the case were it not for his absence, sickness or other disability.

SEC. 5. In a sparce population, a physician, when visiting a sick person, may be desired to see, in an emergency, a neighboring patient who is under the regular charge of another physician. The conduct to be pursued on such an occasion is: to give advice adapted to present circumstances; to interfere as little as possible with the general plan of treatment; to assume no farther directions of the case unless it be expressly desired; and, in the latter case, to request an immediate consultation with the practitioner previously employed.

SEC. 6. A wealthy physician should not give advice *gratis* to the affluent; because his so doing is an injury to his professional brethren. The office of the physician can never be supported as an exclusively beneficent one; and it is defrauding, in some degree, the common fund when fees are dispensed with which might justly be claimed.

SEC. 7. When a physician who has been engaged to attend a case of midwifery is absent and another is sent for, if delivery is accomplished in the absence of the former, the latter is entitled to the fee, but he should resign the patient to the practitioner first engaged.

Article VI.—Of Differences Between Physicians.

SECTION 1. Diversity of opinion and opposition of interests may, in the medical, as in other professions, sometimes occasion controversy and even contention. When such cases occur and cannot be immediately terminated, they should be referred to the arbitration of a sufficient number of physicians, or a *court-medical*.

Article VII. Of Pecuniary Acknowledgments.

SECTION 1. Some general rules should be adopted by the physicians in every town or district, relative to pecuniary acknowledgments from patients. These should be adhered to by physicians as uniformly as circumstances will permit. They serve, likewise, as a standard to which appeal may be taken in cases of doubt or dispute.

SEC. 2. Members of the medical profession have been so uniformly in the habit of attending, gratuitously, the indigent sick, and, in general, of answering every call promptly, and without a question as to whether they are to receive remuneration therefor, that many persons seem to think they have a right to demand the services of a physician; and do, in fact, call upon them freely, and neglect or refuse to render any pecuniary equivalent, although abundantly able to do so. They impose upon one physician, in this way, until they have exhausted his patience, and then call upon another; and thus, in the course of a few years, make the circuit of the profession in their neighborhood. It is proper for the physicians of a community to make a list of the names of such individuals, and to demand, before visiting those whose names are on it, adequate security that their *honorarium* will be paid.

PART III.

THE RECIPROCAL DUTIES AND OBLIGATIONS OF PHYSICIANS AND THE PUBLIC.

Article I. Duties of Physicians to the Public.

SECTION 1. As good citizens, it is the duty of physicians to be vigilant for the welfare of the community, and to bear their part in sustaining its institutions and burdens. They should be always ready to give counsel to the public, in relation to matters appertaining to their profession: as, for example, on subjects of medical police, public hygiene and legal medicine. It is their province to enlighten the public in regard to quarantine regulations, the location, arrangement and dietaries of hospitals, asylums, schools, prisons and similar institutions; in relation to the medical police of towns, drainage, ventilation, etc., and in regard to measures for the prevention of epidemic and contagious diseases. And, when pestilence prevails, it is their duty to face the danger, and to continue their labors for the alleviation of suffering, and the saving of life, even at the risk of their own lives.

SEC. 2. Physicians should always be ready when called on by the proper authorities, to enlighten coroners' inquests and courts of justice on matters strictly medical, such as involve questions relating to insanity, legitimacy, or sudden and violent deaths, and in regard to the various other subjects embraced in the science of medical jurisprudence. But, in these cases, and especially where they are required to make *post-mortem* examinations, it is just and right, in consequence of the time, labor and skill required, and the responsibility and risk they incur, that the public should award them more than a mere consulting fee.

SEC. 3. There is no profession, by the members of which eleemosynary services are more freely dispensed than they are by physicians; but justice demands that some limit should be placed to the claims upon such offices at their hands. Poverty, professional brotherhood, the benevolent and scantily remunerated occupation of the individual patient, and certain of the public duties referred to in Sec. 1 of this Article, should always be recognized as presenting valid claims for gratuitous services. But neither institutions endowed by the public or by rich individuals, societies for mutual benefit, for the insurance of lives or for analagous purposes, nor any profession or occupation can be admitted to possess such privilege. Nor can it be justly expected of physicians to furnish certificates of inability to serve on juries, or perform military duty, or to certify to the state of health of parties wishing to insure their lives, obtain pensions or the like, without a pecuniary acknowledgment. But to indigent persons such professional services should always be cheerfully and freely accorded.

Article II. Obligations of the Public to Physicians.

SECTION 1. The benefit accruing to the public, directly and indirectly, from the active and constant labors and beneficence of the medical profession are so numerous and important that physicians are justly entitled to the utmost consideration from the community. The public ought, likewise, to entertain a just appreciation of the proper qualifications of a practitioner of medicine; to make a due discrimination between true science and the assumptions of ignorance and empiricism; to afford every encouragement and facility for the acquisition of medical education, and not to allow the provisions of their statute books or of the prospectus of their chartered institutions to interpose any obstacles to the attainment of the fullest knowledge of every branch of medical science, or, in any way, to restrain the most entire freedom of thought, investigation, and action in matters appertaining to the practice of medicine.

CODE OF ETHICS OF THE NATIONAL ECLECTIC MEDICAL ASSOCIATION.

SECTION 1. The members of this association shall exercise toward each other, toward all physicians, eclectics especially, and toward all mankind, that courtesy and just dealing to which every one in his legitimate sphere is entitled, and any departure therefrom shall be deemed unprofessional, undignified and unworthy the honorable practitioner of an honorable profession. It shall be regarded as unbecoming to engage in any form of practice, or of advertising, which shall tend to lower the physician in the esteem of the community, or to reflect discredit upon his professional associates.

SEC. 2. While it is the undoubted right of every physician to present himself before the public in an honorable manner, and to state that he makes a specialty of any particular disease, no member of this association shall advertise himself by handbills, circulars, publication of certificates of cures, or any such means; nor associate himself in business professionally with any one so doing; nor advertise himself as belonging to this association, or any auxiliary medical society or any medical college. Any member knowing of any violation of this provision by members of this association, or of any person not a member of this association or any auxiliary medical society advertising himself as such, shall inform the Executive Committee of the matter with all the facts in his possession; and it shall be the duty of the Executive Committee thereupon to publish the facts in some public journal, circulating in the region where such offense has been committed.

Discipline of Members.

SECTION 1. Any member may be officially censured, invited to withdraw, or expelled from membership, for improper conduct, or a violation of professional comity. But it shall be necessary for a specific charge to be made in writing, and a copy to be presented to the person accused, or some person acting in his behalf, and another placed in the hands of the President or Secretary one month before the time of holding a regular meeting.

Sec. 2. All professors or officers of colleges voting and otherwise co-operating in the conferring of the degree of Doctor of Medicine on any person not duly entitled to the same, by the necessary attendance on medical lectures and thorough examinations, shall be considered as liable to the penalties enumerated in this article.

CHICAGO MEDICAL SOCIETY.

FEE BILL, ADOPTED MARCH, 1875.

Inasmuch as the following table of charges is founded upon a just considera-tion of the services performed by medical practitioners, it will be considered a duty on their part to conform to it, whenever the circumstances of their patients do not clearly forbid such a course.

MEDICINE.

For an ordinary visit in the day time	$ 2 00 to	$ 5 00
For each additional person prescribed for, when more than one member of the family is sick at the same time	1 00 to	3 00
For a visit made between 10 p. m. and 6 a. m.	5 00 to	10 00
For rising at night	2 00 to	3 00
For a visit as consulting physician	5 00 to	15 00
For each subsequent visit as consulting physician, in the same case	3 00 to	10 00
For unusual detention, per hour	2 00 to	5 00
For office consultation, according to importance of case	1 00 to	10 00
For a letter of advice, or written opinion	5 00 to	25 00
For examination for life insurance	3 00 to	5 00
For certificate as family medical attendant	2 00 to	3 00
For an examination involving a question of law, in a case in which the physician may be subpœned	10 00 to	100 00
For a visit to a small-pox patient, an additional fee of	1 00 to	3 00
For a post-mortem examination in a case of legal investigation	50 00 to	100 00
For attendance upon court, per day	50 00 to	100 00
For services to distant patients, in addition to expenses of travel, per day	50 00 to	100 00

OBSTETRICS.

For attendance upon a case of natural delivery	$ 10 00 to	100 00
For cases complicated by hemorrhage, convulsions, or other causes, involving extra care or responsibility	25 00 to	100 00
For obstetrical operations, as turning, application of forceps, craniotomy, cephalo-tripsy, Cæsarian section, etc	25 00 to	200 00
For topical treatment of uterine diseases, each time	3 00 to	10 00
For operation of uterine tumors	25 00 to	100 00
For vesico-vaginal or vesico-rectal fistula	50 00 to	300 00
For ovariotomy	100 00 to	500 00

SURGERY.

For amputation of the thigh	$ 75 00 to	300 00
For amputation at knee	50 00 to	300 00
For amputation of the leg	50 00 to	200 00
For amputation of the foot	50 00 to	100 00
For amputation of fingers or toes	10 00 to	25 00
For amputation at shoulder joint	75 00 to	200 00
For amputation of the arm	50 00 to	150 00
For amputation of forearm or hand	50 00 to	100 00
For resection of head of femur	75 00 to	300 00
For resection of knee	75 00 to	200 00
For resection of the shoulder	50 00 to	200 00

For resection of the elbow	50 00 to 100 00
For operations for necrosis	25 00 to 100 00
For reducing dislocation of hip	50 00 to 200 00
For reducing dislocation of the knee	50 00 to 100 00
For reducing dislocation of the ankle	50 00 to 75 00
For reducing dislocation of the fingers or toes	10 00 to 20 00
For reducing dislocation of shoulder or elbow	25 00 to 50 00
For reducing dislocation of wrist	15 00 to 50 00
For reducing dislocation of jaw	20 00 to 50 00
For reducing fracture of the femur	25 00 to 50 00
For reducing fracture of the leg	20 00 to 50 00
For reducing fracture of ribs	5 00 to 20 00
For reducing fracture of the arm or forearm	20 00 to 50 00
For reducing fracture of the small bones	5 00 to 10 00
For reducing fracture of the clavicle	10 00 to 25 00
For operation for internal piles	25 00 to 100 00
For operation for external piles	10 00 to 25 00
For operation for polypi of rectum, nose or ear	20 00 to 50 00
For operation for anal fissure	10 00 to 50 00
For operation for lacerated perinaeum	20 00 to 100 00
For operation for fistula-in-ano	25 00 to 50 00
For operation for radical cure of hernia	30 00 to 100 00
For operation for stone in the bladder	75 00 to 500 00
For operation for cataract	50 00 to 200 00
For operation for strabismus	25 00 to 50 00
For operation on the eyelids	10 00 to 50 00
For operation for artificial pupil	20 00 to 100 00
For operation for radical cure of hydrocele	20 00 to 50 00
For tapping hydrocele	5 00 to 20 00
For operation for varicocele	25 00 to 50 00
For operation for varicose veins	15 00 to 50 00
For reduction of strangulated hernia by taxis	10 00 to 25 00
For performance of tracheotomy	100 00 to 200 00
For pneumatic aspiration	50 00 to 200 00
For operation for club-foot	25 00 to 200 00
For treatment for ununited fracture	25 00 to 200 00
For trephining cranium	50 00 to 200 00
For tapping abdomen	10 00 to 25 00
For operation for cleft palate	25 00 to 100 00
For dilating stricture urethra, each operation	5 00 to 10 00
For introduction of catheter	3 00 to 10 00
For vaccination	1 00 to 5 00

In all cases of gonorrhœa and syphilis, a fee of from $10.00 to $20.00 will be required in advance, the subsequent charges being graduated according to amount of after-attendance necessary.

The foregoing charges are for the performance of the operation only. The subsequent visits are to be charged as in attendance in ordinary cases of disease, the fee being proportioned always to the time occupied and the trouble and responsibility incurred.

For operations and services not enumerated in the foregoing list, charges will be made according to their nature, extent and importance.

PUBLIC INSTITUTIONS.

UNITED STATES MARINE HOSPITAL.

The United States Marine Hospital for the port of Chicago is very pleasantly located at Lake View and has excellent accommodations at the present time for 300 patients—the medical staff at this date is as follows:

H. W. Austin (In charge), Surgeon United States Marine Hospital Service. G. M. Magruder, J. O. Cobb, A. W. Condict, Assistant Surgeons United States Marine Hospital Service.

NORTHERN HOSPITAL FOR THE INSANE.

LOCATED AT ELGIN.

Superintendent...E. A. Kilbourne, M.D.

EASTERN HOSPITAL FOR THE INSANE.

LOCATED AT KANKAKEE.

Superintendent...R. S. Dewey, M.D.

CENTRAL HOSPITAL FOR THE INSANE.

LOCATED AT JACKSONVILLE.

Superintendent...Henry F. Carriel, M.D.

SOUTHERN HOSPITAL FOR THE INSANE.

LOCATED AT ANNA.

Superintendent...Horace Wardner, M.D.

INSTITUTION FOR THE DEAF AND DUMB.

LOCATED AT JACKSONVILLE.

Superintendent...Philip G. Gillett, LL.D.

INSTITUTION FOR THE BLIND.

LOCATED AT JACKSONVILLE.

Superintendent...Rev. F. W. Phillips.

ASYLUM FOR FEEBLE-MINDED CHILDREN.

LOCATED AT LINCOLN.

Superintendent...W. B. Fish, M.D.

SOLDIERS' ORPHANS' HOME.

LOCATED AT NORMAL.

Superintendent...Dr. H. C. DeMatte.

ILLINOIS CHARITABLE EYE AND EAR INFIRMARY.

Chicago, corner Adams and Peoria streets. Founded, May, 1858, as the Chicago Charitable Eye and Ear Infirmiry. Chartered in 1865. Transferred to the State authorities in 1871. The object of the Infirmary is to provide gratuitous medical and surgical treatment for the poor suffering with diseases of the eye or ear.

There is in connection with this Infirmary a Dispensary Department for the treatment of "out patients," which is open from 1.30 to 3 P.M. daily, except Sunday and legal holidays.

TRUSTEES.—Pres., Dan'l Goodwin, Jr.; Vice-Pres., E. S. Fowler, M. D.: Sec. W. H. Fitch, M. D.; Treas., W. I. Culver: Supt., E. C. Lawton.

EYE DEPARTMENT.—Senior Surgeon, E. L. Holmes; Surgeons Eye Department, F. C. Hotz, Lyman Ware, W. T. Montgomery, E. J. Gardiner, Assistant Surgeons, A. P. Gilmore, J. E. Colburn, G. F. Fiske, B. Bettman, C. E. Walker, C. H. Beard, G. E. Brinkerhoff.

EAR DEPARTMENT.—Surgeons, S. S. Bishop, I. E. Marshall; Assistant Surgeons, J. J. Alderson, J. R. Davey, Chas. Davison, G. W. Webster; Microscopist, I. N. Danforth; Resident Assistant Surgeon, Wm. L. Noble.

COOK COUNTY HOSPITAL.

(Telephone 7133.)

Corner of Wood and Harrison streets. Organized in 1865. Maintained and controlled by the Board of County Commissioners.

MEDICAL BOARDS.

Regular Medical Board.—Surgeons Christian Fenger, D. A. K. Steele, A. E. Hoadley, Albert R. Strong, Oscar J. Price, A. J. Baxter, W. T. Belfield, D. W. Graham. Physicians—J. P. Ross, P. J. Rowan, J. R. Brandt, A. J. Corey, A. C. Cotton, E. V. McDonald, E. Melms, J. A. Robison.

Gynæcologists.—John Guerin, F. Henrotin, D. Servis, Chas. W. Earle.

Oculist and Aurist.—J. Elliott Coburn.

Pathologist.—Elbert Wing.

Homœopathic Medical Board.—A. W. Burnside, Chas. Gatchell, N. B. Delamater, A. W. Woodward, E. H. Pratt, Henry Sherry.

Gynæcologist.—John W. Streeter.

COOK COUNTY POOR HOUSE.

Telephone 4334.

Is located twelve miles from the city, on the County Poor Farm, in Norwood Park Township, on which is also the Cook County Insane Asylum. there are nine buildings. They were completed in 1882, and have a capacity of 1,000 patients.

Medical Staff.—Medical Superintendent, T. J. Conley; Physician, E. E. Schenault.

COOK COUNTY INSANE ASYLUM.

Telephone 4334.

Telephone 4334. Is located on the County Poor Farm.

Medical Staff.—Medical Superintendent, Dr. John C. Spray; Assistant Physicians, Dr. Frank V. Luse, Dr. Florence W. Hunt, Dr. Oscan Bluthardt.

COOK COUNTY MORGUE

is situated on the County Hospital grounds, at the rear of the hospital.

COOK COUNTY.

County Physician, H. N. Moyer;
Ass't. County Physician, J. L. Gray.

CITY OF CHICAGO.

DEPARTMENT OF HEALTH.

Office, City Hall, LaSalle between Randolph and Washington Streets.

Commissioner of Health.—O. C. DeWolf, M.D.; Ass't Comr. and Dept. Secy., H. P. Thompson.

Medical Inspector, South Division.—Frank Cary, M.D.

Medical Inspector, North Division.—Junius M. Hall, M.D.

Medical Inspector Southwest Division.—Erasmus Garrott, M.D.

Medical Inspector Northwest Division.—Liston H. Montgomery, M.D.

There are twenty-four Sanitary Policemen and twenty-four Tenement and Factory Inspectors under the charge of the Commissioner of Health.

Small-Pox Hospital, corner Twenty-sixth and California Avenue.

Patients are sent to Hospital under the direction of the Commissioner of Health. The Hospital is in charge of (Sisters) order of "Poor Handmaids of Jesus."

City Physician.—Appointed biennially by the Mayor and City Council. Salary, $2,000 per annum. His principal duties are to attend the police stations and the Bridewell. Present incumbent, A. L. Corey.

SCHOOLS AND COLLEGES.

CHICAGO MEDICAL COLLEGE.

MEDICAL DEPARTMENT NORTHWESTERN UNIVERSITY,

PRAIRIE AVENUE, COR. 26TH STREET.

(Telephone 8128.)

Faculty.--Emeritus Professor of the Principles and Practice of Medicine and of Clinical Medicine, H. A. Johnson, M.D., LL.D.; Emeritus Professor of Obstetrics, E. O. F. Roler, A.M., M.D.; Professor of Principles and Practice of Medicine and of Clinical Medicine, N. S. Davis, M.D., LL.D., Dean; Professor of Clinical Surgery, Edmund Andrews, M.D., LL.D., Treasurer; Professor of the Principles and Practice of Surgery and of Clinical Surgery, Ralph N. Isham, A.M., M.D.; Professor of Clinical Medicine, John H. Hollister, A.M., M.D.; Professor of Ophthalmology and Otology, S. J. Jones, M.D., LL.D.; Professor of Diseases of Children, M. P. Hatfield, A.M., M.D.; Professor of General and Medical Chemistry, John H. Long, Sc.D.; Professor of Gynæcology, E. C. Dudley, A.B., M.D.; Professor of Surgical Anatomy and Operative Surgery, and of Clinical Surgery, John E. Owens, M.D.; Professor of State Medicine and Public Hygiene, O. C. DeWolf, A.M., M.D.; Professor of Nervous and Mental Diseases, and of Medical Jurisprudence, Walter Hay, M.D., LL.D.; Professor of Descriptive and Surgical Anatomy, F. C. Schaefer, M.D.; Professor of Clinical Medicine, I. N. Danforth, A.M., M.D.; Professor of Materia Medica and Therapeutics, and of Laryngology and Rhinology, W. E. Casselberry, M.D.; Professor of Obstetrics, W. W. Jaggard, A.M., M.D.; Professor of Principles and Practice of Medicine, N. S. Davis, Jr., A.M., M.D.; Professor of General Pathology and Pathological Anatomy, F. S. Johnson, A.M., M.D.; Professor of Physical Diagnosis and Clinical Medicine, Frank Billings, M.D., Secretary; Professor of Clinical Surgery. E. Wyllys Andrews, A.M., M.D.; Professor of Histology, Frank T. Andrews, A.M., M.D.

Lecturers.—Lecturer on Physiology, George W. Webster, M.D.; Demonstrator of Pathology, Elbert Wing, A.M., M.D.; Demonstrator of Operative Surgery, William E. Morgan, M.D.; Demonstrator of Anatomy, Herbert H. Frothingham, M.D.; Demonstrator of Histology, J. C. Hoag, Ph.M., M.D.; Clinical Assistant to the Professor of Surgery, George S. Isham, A.M., M.D.; Assistant Demonstrator of Anatomy, W. N. Hibbard, A.M., M.D.; J. O. Wakem, Clerk.

College year commences Tuesday, September 25, 1888, and terminates Tuesday, March 26, 1889.

RUSH MEDICAL COLLEGE.

MEDICAL DEPARTMENT LAKE FOREST UNIVERSITY.

(Telephone 7113.)

Corner of Wood and Harrison Streets. Incorporated, 1837. Organized, 1843. The winter term begins the third Tuesday in September, and closes the second Tuesday of February of the following year. The spring term opens the third Tuesday in February, and closes the second Tuesday in June.

Board of Trustees.—Hon. L. C. P. Freer, President; Joseph P. Ross, M.D., Vice President; Hon. Grant Goodrich, LL.D., Secretary; Chas. T. Parkes, M.D., Treasurer; Henry M. Lyman, M.D.; Assistant Secretary; J. Adams Allen, M.D.; DeLaskie Miller, M.D.; Charles T. Parkes, M.D.; Edward L. Holmes, M.D.; James H. Etheridge, M.D.; W. S. Haines, M.D.; Hon. John C. Haines; Rev. Wm. C. Roberts, D.D., President Lake Forest University; His Excellency, Richard J. Oglesby, Governor; The speaker of the House of Representatives, J. Adams Allen, M.D., LL.D., President of the College, *Ex-officio.*

Faculty.—J. Adams Allen, M.D., LL.D., President, Professor of the Principles and Practice of Medicine, 125 State Street; Joseph P. Ross, A.M., M.D., Professor of Clinical Medicine and Diseases of the Chest, 428 Washington Boulevard; William H. Byford, A.M., M.D., Professor of Gynæcology, 125 State Street; Edward L. Holmes, A.M., M.D., Professor of Diseases of the Eye and Ear, 112 Clark Street; Henry M. Lyman, A.M., M.D., Professor of Physiology and of Diseases of the Nervous System, 69 Dearborn Street; James H. Etheridge, A.M., M.D., Secretary; Professor of Materia Medica and Medical Jurisprudence, 1634 Michigan Avenue; Charles T. Parkes, M.D., Professor of Principles and Practice of Surgery and Clinical Surgery, 51 Lincoln Avenue; Walter S. Haines, A.M., M.D., Professor of Chemistry, Pharmacy and Toxicology, Laboratory in College Building; James Nevins Hyde, A.M., M.D., Professor of Skin and Venereal Diseases, 240 Wabash Avenue; Norman Bridge, M.D., Professor of Pathology and Adjunct Professor of the Principles and Practice of Medicine, 550 West Jackson Street; Arthur D. Broon, M.D., Professor of Anatomy; Nicholas Senn, M.D., Professor of Surgical Pathology and Principles of Surgery; J. Suydam Knox, A.M., M.D., Professor of Obstetrics.

Professors of Special Departments.—Truman W. Brophy, M.D., D.D.S., Professor of Dental Pathology and Surgery, 125 State Street; E. Fletcher Ingals, A.M., M.D., Professor of Laryngology, 64 State Street.

Adjunct Professors and Lecturers.—Daniel T. Nelson, A.M., M.D., Adjunct Professor of Gynæcology, 125 State Street; Philip Adolphus, M.D., Clinical Adjunct to the chair of Gynæcology, 638 Washington Boulevard; Eugene W. Whitney, A.B., M.D., Lecturer on Surgery, 174 Warren Avenue; Alfred C. Cotton, M.D., Lecturer on Therapeutics, 193 South Wood Street; Eugene S. Talbot, M.D., D.D.S., Lecturer on Dental Anatomy and Physiology, 125 State Street; Samuel J. Holmes, M.D., Lecturer on Pathological Anatomy and Pathological Histology, 315 Fulton Street; Daniel R. Brower, M.D., Lecturer on the Practice of Medicine, 65 Randolph Street; Henry P. Merriman, A.M., M.D., Lecturer on Gynæcology, 2237 Michigan Avenue; John A. Robison, A.M., M.D., Lecturer on Materia Medica, 428 Washington Boulevard; Henry T. Byford, M. D. Lecturer on Obstetrics; Frederick E. Sherman, M.D., Assistant Demonstrator of Anatomy; Wm. H. Morgan, M.D., Clinical Assistant to the Chair of Diseases of Children; Thomas J. Shaw, M.D., Clinical Assistant to the Chair of Gynæcology; A. E. Kauffman, M.D., Demonstrator of Chemistry; Albert J. Ochsner, B.S., M.D., Demonstrator of Physiology and Pathology; Harold N. Moyer, M.D., Assistant to the Chair of Diseases of the Nervous System; Cassius D. Wescott, Demonstrator of Anatomy; Mr. Frank Jordan Gould, College Clerk.

COLLEGE OF PHYSICIANS AND SURGEONS.

Organized 1881, 813 W. Harrison St.

Telephone 7121.

Board of Directors.—President, A. Reeves Jackson, M.D; Vice-President, S. A. McWilliams, M.D.; Secretary, D. A. K. Steele, M.D.; Treasurer, Leonard St. John, M.D.; Chairman Finance Committee, Charles Warrington Earle, M.D.

Faculty.—Professor of Gynæcology.—A. Reeves Jackson, A.M., M.D., 271 Michigan avenue. Professor of Clinical Medicine, Diseases of the Chest, and Physical Diagnosis.—Samuel A. McWilliams, A.M., M.D., 3456 Michigan avenue. Professor of Principles and Practices of Surgery and Clinical Surgery.—Daniel A. K. Steele, M.D., 1801 State street. Professor of Demonstrations of Surgery and Surgical Appliances.—Leonard St. John, M.D., C. M., M. R. C. S., Eng., 539 West Monroe street. Professor of Obstetrics. Chas. Warrington Earle, A.M., M.D., 535 Washington Boulevard. Professor of Operative Surgery, Clinical Surgery, and Surgical Pathology. Henry Palmer, M.D., Janesville, Wisconsin. Professor of Otology, Laryngology and Rhinology.—Frank E. Waxham, M.D., 243 State street. Professor of Ophthalmology and Otology.—John E. Harper, A.M., M.D., 163 State street. Professor of Dental Surgery.—A. W. Harlan, M.D., D. D. S., 70 Dearborn street. Professor of Orthopedic Surgery, Albert E. Hoadley, M.D., 683 Washington Boulevard. Professor of Therapeutics.—C. C. P. Silva, M.D., 98 Ogden avenue. Professor of Diseases of the Mind and Nervous System.—Oscar A. King, M.D., 70 Monroe street. Professor of General Pathology, Hygiene and Bacteriology.—Romaine J. Curtiss, M.D., Joliet, Illinois. Professor of Medical Chemistry.—W. K. Harrison, A.B., M.D., 295 North State street. Professor of Practice of Medicine and Clinical Medicine.—Wm. E. Quine, M.D., 3160 Indiana avenue. Professor of Surgical Diseases of the Genito-Urinary System.—J. T. Jelks, M.D., Hot Springs, Ark. Professor of Dermatology.—Henry J. Reynolds, M.D., 163 State street. Professor of Clinical Surgery.—Christian Fenger, M.D., 125 State street. Professor of Inorganic Chemistry.—Charles B. Gibson, M.D., 81 Clark Street; Professor of Physiology, John A. Benson, A.M., M.D., 163 State street. Professor of Diseases of Children.—H. P. Newman, M.D.

LECTURERS IN THE SPRING COURSE.

Lecturer on Gynæcology.—E. P. Murdock, A.M., M.D., 148 Loomis street Lecturer on Obstetrics, H. P. Newman, M.D., 65 Randolph street. Lecturer on Surgical Diseases of the Genito-Urinary System, G. Frank Lydston, M.D., 112 Clark street. Lecturer on Anatomy.—B. A. Brigham, M.D. Lecturer on Histology and Microscopy.—A. J. C. Saunier, M.D., Libertyville, Ill. Lecturer on Surgical Anatomy, Elmer E. Babcock, M.D., 1801 State street. Lecturer on Materia Medica.—W. C. Caldwell, M.D., 168 Halsted street. Lecturer on Ophthalmology and Otology.—G. Erwin Brinckerhoff, M.D., 227 West Adams street. Lecturer on Surgery.—Thos. A. Davis, M.D., 975 West Van Buren street. Lecturer on Practice of Medicine.—F. B. Earle, M.D. Demonstrator of Anatomy.—Wm. Goodsmith, M.D. Clerk.—Levi Dell.

WOMAN'S MEDICAL COLLEGE.

335 S. LINCOLN ST.

(Telephone 7116.)

Board of Trustees. Term Expires 1888.—President, Wm. H. Byford, A.M., M.D.; Secretary and Treasurer, Chas. W. Earle, A.M., M.D.; E. Fletcher

Ingals, A.M., M.D. Term expires 1889.—D. W. Graham, A.M., M.D.; I. N. Danforth, A.M., M.D.; Wm. J. Maynard, A.M., M.D. Term expires 1890.—Sarah Hackett Stevenson, M.D.; Daniel R. Brower, M.D.; F. L. Wadsworth, M.D.

Emeritus Faculty.—Professor Emeritus of Theory and Practice of Medicine, W. Godfrey Dyas, M.D., F.R.C.S.; Professor Emeritus of Materia Medica and Therapeutics, G. C. Paoli, M.D.; Professor Emeritus of Gynæcology, T. Davis Fitch, M.D.; Professor Emeritus of Surgery, R. G. Bogue, M.D.

Faculty.—Professor of Gynæcology, William H. Byford, A.M., M.D., President; Professor of Diseases of Children and Clinical Medicine, Charles Warrington Earle, A.M., M.D., Treasurer; Professor of Renal Diseases, Isaac N. Danforth, A.M., M.D.; Professor of Theory and Practice of Medicine, Henry M. Lyman, A.M., M.D.; Professor of Diseases of the Nervous System and Clinical Medicine, Daniel R. Brower, M.D.; Professor of Obstetrics, Sarah Hackett Stevenson, M.D.; Professor of Surgery, David W. Graham, A.M., M.D.; Professor of Dermatology, Wm. J. Maynard, A.M., M.D.; Professor of Ophthalmology and Otology, Wm. T. Montgomery, M.D.; Professor of Diseases of the Chest and Throat, E. Fletcher Ingals, A.M., M.D.; Professor of Physiology, F. L. Wadsworth, M.D.; Professor of Gynæcology, Marie J. Mergler, M.D., Secretary; Professor of Dental Surgery, Eugene S. Talbot, M.D., D.D.S.; Professor of Chemistry and Toxicology, Jerome H. Salisbury, A.B., M.D.; Professor of Anatomy, Mary E. Bates, M.D.; Professor of Materia Medica and Therapeutics, John A. Robison, M.D.; Clinical Professor of Gynæcology at the Hospital for Women and Children, Mary H. Thompson, M.D.; Professor of Hygiene and Medical Jurisprudence, Eliza H. Root, M.D.; Professor of Pathology and Director of the Pathological Laboratory, Frank Cary, M.D.

Lecturers and Assistants.—Clinical Lecturer on Midwifery, and in charge of outside obstetrical department, Robert S. Hall, M.D.; Lecturer on Diseases of the Chest and Throat, Homer M. Thomas, M.D.; Assistants to the Chair of Anatomy, Annette S. Dobbin, M.D., Ellen Heise, M.D., Rachel Hickey, M.D.; Lecturer on Histology and Director of the Histological Laboratory, Mary Mixer, M.D.; Assistant to the Chair of Physiology, Deroxa N. Morey, B.S., M.D.; College Clerk and Janitor, M. L. Bush.

THE HAHNEMANN MEDICAL COLLEGE OF CHICAGO.
2813 COTTAGE GROVE AVENUE.

Incorporated 1855. Organized 1859.

Officers of the College.—President, D. S. Smith, M.D.; Vice-President, Erskine M. Phelps; Secretary, G. F. Shears, M.D.; Treasurer, Temple S. Hoyne, M.D.; Trustees, D. S. Smith, M.D.; E. H. Sheldon, Esq.; J. Young Scammon, LL. D.; R. L. North, Esq.; T. S. Hoyne, M.D.; E. M. Phelps, Esq.; H. J. MacFarland, Esq.; H. N. Higinbotham, Esq.; Byron L. Smith, Esq.; R. R. Cable, Esq.

Officers of the Faculty.—President, D. S. Smith, M.D.; Dean, R. Ludlam, M.D.; Registrar, E. S. Bailey, M.D.

Faculty.—Emeritus Professor of Materia Medica and Therapeutics, D. S. Smith, M.D.; Professors of the Principles and Practice of Medicine, *H. B. Fellows, M.D., 2969 Indiana Avenue; *C. E. Laning, M.D., Central Music Hall; Professors of the Medical and Surgical Diseases of Women, and Obstetrics, *R. Ludlam, M.D., 1823 Michigan Avenue; S. Leavitt, M.D., 148 Thirty-seventh Street; E. S. Bailey, M.D., 3034 Michigan Avenue; Professor of the Principles and Prac-

*The star indicates the governing faculty.

tice of Surgery, *Geo. F. Shears, M.D., 3130 Indiana Avenue; Professor of Materia Medica and Therapeutics, *W. J. Hawkes, M.D., 241 Dearborn Avenue; Professor of Diseases of the Eye and Ear, *C. H. Vilas, M.A., M.D., Central Music Hall; Professor of Chemistry and Toxicology, *J. B. S. King, M.D., 240 Wabash Avenue; Professor of Descriptive and Practical Anatomy, C. E. Laning, M.D., Central Music Hall; Professor of Physiology, Histology and Sanitary Science, *J. E. Gilman, M.D., 455 W. Washington Street: Adjunct Professor of Chemistry and Toxicology, J. Eugene Jordan, M.D.; Adjunct Professor of Physiology and Histology, J. P. Cobb, M.D.; Adjunct Professor of Materia Medica, W. S. Gee, M.D.: Adjunct Professor of the Diseases of the Eye and the Ear, Joseph Watry, M.D.: Adjunct to the Chair of Theory and Practice, B. S. Arnulphy, M.D.: Adjunct to the Chair of Anatomy, Homer V. Halbert, M.D.; Demonstrator of Anatomy, O. M. Baird, M.D.

CHICAGO HOMŒOPATHIC MEDICAL COLLEGE.

CORNER WOOD AND YORK STREETS.

Chartered 1876.

Board of Counsellors.—Geo. E. Shipman, M.D., Leonard Pratt, M.D., Willis Danforth, M.D., E. M. Hale, M.D., Charles Adams, M.D., M. B. Campbell, M.D., C. C. Bonney, L.L.D., Hon. W. H. Wood, P. B. Weare, Esq., Henry Strong., Esq., Hon. W. C. Goudy, Edson Keith, Esq., Hon. J. Russell Jones, Marvin Hughitt, Esq., Philo R. King, Esq., Francis B. Peabody.

Officers of the College.—President, J. S. Mitchell, M.D.; Vice-President, A. W. Woodward, M.D.; Secretary, J. R. Kippax, M.D.; Treasurer, L. C. Grosvenor, M.D.; Manager, J. H. Buffum, M.D.

Regular Faculty. Theory and Practice of Medicine.—J. S. Mitchell, A.M., M.D., 2432 Michigan Avenue, Professor of Institutes and Practice of Medicine: John R. Kippax, M.D., LL. B., 3154 Indiana Avenue, Professor of Principles and Practice of Medicine and Medical Jurisprudence; N.B. Delamater, A.M., M.D., 125 State Street, Professor of Mental and Nervous Diseases; R. N. Tooker, M.D., 261 Dearborn Avenue, Professor of Diseases of Children. Surgery.—Albert G. Beebe, A.M., M.D., 552 West Monroe Street, Professor of Operative Surgery; E. H. Pratt, M.D. LL.D., Central Music Hall, Professor of Principles and Practice of Surgery; W. F. Knoll, M.D., 726 Washington Boulevard, Professor of Minor Surgery; F. H. Gardiner, M.D., D.D.S., 126 State Street, Lecturer on Dental Surgery. Obstetrics and Gynæcology.—R. N. Foster, A.M., M.D., 10 Warren Avenue, Professor of Obstetrics; John W. Streeter, M.D., 2001 Prairie Avenue, Professor of Medical and Surgical Diseases of Women. Ophthalmology and Otology.—J. H. Buffum, M.D., 100 State Street, Professor of Diseases of the Eye and Ear; W. M. Stearns, M.D., 100 State Street. Materia Medica and Therapeutics.—A. W. Woodward, M.D., 130 Ashland Avenue, Professor of Materia Medica and Therapeutics; H. M. Hobart, A.M., M.D., 402 Center Street, Professor of Materia Medica. Anatomy.—Curtis M. Beebe, M.D., 27 Ashland Avenue, Professor of Anatomy. Physiology, Pathology, Histology and Microscopy.—W. F. Knoll M.D., 726 Washington Boulevard, Professor of Physiology, Pathology and Minor Surgery; M. J. Bliem, Ph.B., M.D., 308 Warren Avenue, Adjunct Professor of Physiology and Demonstrator of Histology and Microscopy. Chemistry, Toxicology and Sanitary Science.—Clifford Mitchell, A.M., M.D., 44 Sixteenth Street, Professor of Chemistry and Toxicology; L. C. Grosvenor, M.D., 185 Lincoln Avenue, Professor of Sanitary Science.

BENNETT MEDICAL COLLEGE.

(Telephone 667.)

Nos. 511 and 513 State street. Chartered March 25th, 1869.

The regular term commences in September, and continues six months.

Board of Trustees.--Pres., A. L. Clark; Vice-Pres., Edgar Reading; Sec., Milton Jay; Treas., H. T. Clark; J. B. McFatrich, H. K. Whitford, E. M. Reading, H K. Stratford, Finley Ellingwood, M. G. Hart, H. S. Tucker, E. F. Buecking, John Tascher.

Faculty.--Milton Jay, Dean, Principles and Practice of Surgery and Clinical Surgery; Anson L. Clark, Obstetrics and Diseases of Women and Clinical Gynæcology; L. D. Batchelder, Principles and Practice of Medicine; H. K. Whitford, Clinical Medicine: E. M. Reading, Diseases of the Respiratory and Circulatory Organs and the Nervous System: H. S. Tucker, Descriptive Anatomy: A. L. Willard, Morbid Anatomy and Astrology; Finley Ellingwood, Hygiene and State Medicine, and Analytical Chemistry. E. F. Buecking, Surgical Anatomy and Orthopædic Surgery: J. B. McFatrich, Didactic and Clinical Ophthalmology and Otology; Martin G. Hart, Physiology; Gilbert Wheeler, Chemistry, Pharmacy and Toxicology; John Tascher, Diseases of Children; Edgar Whitford, Dermatology and Venereal Diseases; George C. Christian, Medical Jurisprudence; Eli Wight, Professor of Obstetrics; O. O. Baines, Demonstrator of Anatomy; C. E. Starret, Therapeutics; J. M. Jacobs Materia Medica.

CHICAGO COLLEGE OF DENTAL SURGERY.

NORTHEAST CORNER OF WABASH AVENUE AND MADISON STREET.

Board of Counselors.-Hon. Lyman Trumbull, Hon. John Wentworth, Hon. Carlile Mason, Hon. B. C. Cook, Hon. Henry M. Shepard, Hon. Carter H. Harrison, Wm. K. Ackerman, Esq.; Hon. J. C. Knickerbocker, Dr. G. F. Root, Murry Nelson, Esq.; Wm. Penn Nixon, Esq; John V. Farwell, Esq.; Wm. M. Hoyt, Esq.; Geo. M. Pullman, Esq.; Hon. John A. Roche; E. H. Sargent, Esq.; Wirt Dexter, Esq.; Sidney C. Eastman, Esq.

Board of Directors.—President and Dean of the Faculty, Truman W. Brophy, M.D., D.D.S.; Vice-President, A. W. Harlan, M.D., D.D.S.; Secretary and Treasurer, Edgar D. Swain, D.D.S.; James A. Swasey, N. B. Delamater, A.M., M.D.; E. Fletcher Ingals, A.M., M.D.; A. Reeves Jackson, A.M., M.D.; Milton Jay M.D.; C. H. Thayer, D.D.S.; J. H. Etheridge, A.M., M.D.; G. W. Nichols, M.D.; J. N. Crouse, D.D.S.; Chairman of Executive Committee, Frank H. Gardiner, M.D., D.D.S.

Faculty.—Professor of Operative Dentistry, *Geo. H. Cushing, M.D., D.D.S.; Professor of Pathology, G. V. Black, M.D., D.D.S.; Professor of Anatomy, W. L. Copeland, M.D., C.M., M.R.C.S., England; Professor of Physiology and Microscopy, W. T. Belfield, M.D.; Professor of Oral Surgery, Truman W. Brophy, M.D., D.D.S.; Professor of Materia Medica and Therapeutics, A. W. Harlan, M.D., D.D.S.; Professor of Prosthetic Dentistry, W. B. Ames, D.D.S.; Professor of Regional Anatomy, J. W. Wassall, M.D., D.D.S.; Professor of Irregularities and Hygiene, Garrett Newkirk, M.D.: Professor of Chemistry and Metallurgy,

*Prof. Cushing will be in attendance in the Infirmary and demonstrate Operative Dentistry from 1 to 4 P. M., Mondays and Wednesdays during the entire sessions.

Chas. B. Gibson, M.D.; Adjunct Professor of Operative Dentistry, C. N. Johnson. D.D.S., D.D.S.; Lecturer on Ethics and Dental Jurisprudence, J. M. Crouse, D.D.S.; Superintendent, †G. V. Black, M.D., D.D.S.

Instructors in Dental Technics.—Instructor in Prosthetic Technics, H. P. Smith; Instructor in Operative Technics, D. M. Cattell, D.D.S.

Demonstrators.—Demonstrator of Operative Dentistry, A. H. Peck, D.D.S.; Demonstrator of Microscopy, L. L. Gregory, M.D.; Demonstrator of Chemistry, Prof. Chas. B. Gibson, M.D.; Demonstrator of Anatomy, Wm. H. Weaver, M.D.; First Assistant Demonstrator of Chemistry, C. P. VanGundy, B.Sc.; Assistant Demonstrators of Anatomy, Thos. A. Broadbent, B.S., D.D.S., Chas. J. Merriman, D.D.S.

COLLEGE OF DENTAL AND ORAL SURGERY.

(NORTHWESTERN UNIVERSITY DENTAL COLLEGE.)

College year opens September 25, 1888, closes March 26, 1889.

Faculty.—Joseph Cummings, D.D., LL.D., President; W. W. Allport, M.D., D.D.S., Emeritus Professor of Principles and Practice of Operative Dentistry; *Edmund Andrews, M.D., LL.D., Professor of Clinical Surgery; *Ralph N. Isham, A.M., M.D., Professor of Principles and Practice of Surgery; *Walter Hay, M.D., LL.D., Professor of Medical Jurisprudence; L. P. Haskell, D.D.S., Professor of Prosthetic Dentistry; *F. C. Schaefer, M.D., Professor of Descriptive Anatomy; R. F. Ludwig, D.D.S., Treasurer, Professor of Clinical Operative Dentistry; *John S. Marshall, M.D., Dean, Professor of Oral Surgery; A. E. Baldwin, M.D., D.D.S., Professor of Dental Pathology and Therapeutics; Chas. P. Pruyn, M.D., D.D.S., Professor of Principles and Practice of Operative Dentistry; *John H. Long, Sc.D., Professor of General and Medical Chemistry; Geo. W. Wheaton, M.D., Professor of Physiology; *W. E. Casselberry, M.D., Professor of Materia Medica and Therapeutics; *F. S. Johnson, A.M., M.D., Professor of Pathology and Pathological Anatomy; C. R. Baker, D.D.S., Professor of Metallurgy and Oral Chemistry; Arthur B. Freeman, M.D., D.D.S., Secretary, Professor of Comparative Dental Anatomy; *Frank T. Andrews, A.M., M.D., Professor of Histology; Arthur E. Mattison, Professor of Dental Deformities; J. A. Freeman, Associate Professor of Principles and Practice of Operative Dentistry.

Demonstrators.—*Elbert Wing, A.M., M.D., Demonstrator of Pathology; *Herbert H. Frothingham, M.D., Demonstrator of Anatomy; J. C. Hoog, Ph.M., M.D., Demonstrator of Physiology; David M. Cattell, D.D.S., Demonstrator of Operative Dentistry.

Spring Course.—David M. Cattell, D.D.S., Lecturer on Dental Histology; Herbert H. Frothingham, M.D., Lecturer on Regional Anatomy; W. G. Stowell, D.D.S., Lecturer on Operative Dentistry.

---

†Prof. Black will be in constant attendance supervising the instruction in all of the Clinical Departments.

*Members of the Medical Faculty.

NORTHWESTERN COLLEGE OF DENTAL SURGERY.

DEPARTMENT OF DENTAL AND ORAL SURGERY OF LAKE FOREST UNIVERSITY.

SOUTHEAST COR. WABASH AVE. AND TWELFTH ST.

Incorporated 1885.

Officers.—President, Wm. C. Roberts, D.D., Chancellor of the University; Secretary and Actuary, F. H. B. McDowell.

Board of Directors.—F. H. B. McDowell, N. J. Roberts, Joseph A. Marshall.

Board of Educational Control.—President, J. E. Hequembourg, M.D.; Joseph Haven, M.D.; R. W. Clarkson, D.D.S.; N. J. Roberts, D.D.S.; Byron D. Palmer, D.D.S.

Faculty.—Emeritus Professor of Materia Medica, G. C. Paoli, A.M., M.D.; Emeritus Professor of Pathology, N. P. Pearson, A.M., M.D.; Professor of Operative Dentistry and Dental Histology, R. W. Clarkson, D.D.S.: Professor of Prosthetic Dentistry, Byron D. Palmer, D.D.S.: Professor of Oral Surgery, Norman J. Roberts, D.D.S.; Professor of Physiology and Dean of the Faculty, Jos. Haven, M.D.; Professor of Anatomy and Principles and Practice of Surgery, J. E. Hequembourg, M.D.; Professor of Pathology, J. H. Lyon, A.M., M.D.; Professor of Materia Medica, F. C. Caldwell, M.D.; Professor of Chemistry, J. H. Salisbury, A.M., M.D.

THE AMERICAN COLLEGE OF DENTAL SURGERY.

78-82 STATE ST., CHICAGO.

Board of Directors.—President, T. Davis Fitch, M.D.; Secretary, I. Clendenen, M.D,; Treasurer, L. H. Varney, D.D.S.; Ira E. Marshall, A.M., M.D.; G. C. Varny.

Faculty.—Professor Materia Medica and Therapeutics, T. Davis Fitch, President; Professor Oral Surgery, I. Clendenen, M.D., Secretary, 78 State street; Professor Plastic Surgery, Ward Greene Clark, M.D., Vice-President, S. E. corner Clark and Huron Streets; Professor of Chemistry, W. Meek, A.M., M.D., Secretary of Faculty, 182 Park Avenue; Professor Pathology and Clinical Surgery, Geo. W. Whitefield, M.D., D.D.S., Evanston, Ill.; Professor Physiology, C. T. Hood, A.M., M.D., 255 S. Western Avenue; Professor Electro-Therapeutics and Anæsthetics, E. E. Gwynne, M.D., 407 Centre Street; Professor Anatomy, E. R. Bennett, M.D., 893 Clybourn Avenue; Professor Operative Dentistry, I. B. Crissman. D.D.S., 271 N. Clark Street; Professor Care and Treatment of Children's Teeth and Dental Anatomy, Gustavus North, A.M., D.D.S., Springville, Iowa; Professor Microscopy, L. D. McIntosh, M.D., D.D.S., 141 and 143 Wabash Avenue; fessor Prosthetic Dentistry and Metallurgy, G. A. Thomas, D.D.S., 163 State Street; Emeritus Professor of Chemistry, J. Spafford Hunt, M.D., 878 Fulton Street.

Clinical Instructors.—T. Rin, D.D.S., Dowagiac, Mich., S. M. White, D.D.S., Benton Harbor, Mich., A. J. Holmes, Kalamazoo, Mich., Gustavus North, A.M., D.D.S., Springville, Iowa, Geo. Whitefield, M.D., D.D.S., Illinois, I. B. Crissman, D.D.S., Illinois G. A. Thomas, D.D.S., Illinois, Dr. M. A. Webb, 858 West Van Buren Street, Dr. Charles Upp, 3502 South State Street, L. H. Varney, D.D.S., 78 State Street, Clarke R. Rowley, D.D.S., 163 State Street, E. M. McIntosh, D.D.S., 141 and 143 Wabash Avenue.

THE NORTHWESTERN DENTAL COLLEGE OF CHICAGO

Was organized in the winter of 1887-8, and is very centrally located at 202 State (cor. Adams Street).

The Officers for 1888-9.—President, John Leggett, D.D.S.: President Board of Trustees, Chas. T. McKinney; Secretary, Dr. L. E. Ireland; Treasurer, Dr. A. G. Goodman.

Board of Trustees.—Chas. T. McKinney, Dr. L. E. Ireland, Dr. A. G. Goodman, Dr. John Leggett.

THE ILLINOIS COLLEGE OF PHARMACY.

Department of Pharmacy of Northwestern University, south-west corner Lake and Dearborn Streets, Chicago. Next regular course begins September 27, 1888, and another on the second Thursday in March, 1889.

DEPARTMENTS.—There are four principal departments of study, viz.:

Department of Pharmacy, including Metrology, Pharmaco-technology, Pharmaceutical Chemistry, Pharmacopœias, Pharmaceutical Nomenclature, and Dispensing Pharmacy.

Department of Chemistry, including Natural Philosophy, Chemical and Pharmaceutical Physics, Chemical Philosophy, Inorganic Chemistry, Organic Chemistry, Chemical Analysis, and Toxicological Chemistry.

Department of Pharmacognosy, including Structural and Systematic Botany, Micro-Botany. Principles of Construction of the Microscope, Pharmaceutical Microscopy, and Pharmacognosy.

Department of Materia Medica and Therapeutics, including Physiology, Therapeutics, Inorganic and Organic Meteria Medica, Toxicology and Posology.

FACULTY.—Joseph Cummings, LL. D., President: Oscar Oldberg, P. D., Dean, Professor of Pharmacy and Director of the Pharmaceutical Laboratories: John H. Long, Sc. D., Professor of Chemistry and Director of the Chemical Laboratory; W. E Quine, M. D., Professor of Physiology, Materia Medica and Therapeutics; W. K. Higley, Ph. C., Professor of Botany and Pharmacognosy and Director of the Microscopical Laboratory.

ASSISTANTS.—Mark Powers, B. Sc., Assistant in the Chemical Laboratory: Wm. B. Moore, Ph. G., Assistant to the Chair of Pharmacy, Amanuensis in the Dispensing Laboratory; ——— Assistant in the Microscopical Laboratory.

Occasional lectures are delivered by N. S. Davis, M. D., LL. D.; E. Andrews, M. D., LL. D.; W. H. Byford, A. M., M. D.; A. R. Jackson, A. M., M. D.; Joseph P. Ross, A. M., M. D ; M. P. Hatfield, A. M., M. D.; Charles T. Parkes, M. D.; Chas. W. Earle, A. M. M. D.; J. H. Etheridge, A. M., M. D.; Daniel R. Brower, M. D.; I. N. Danforth, A. M., M. D.; John A. Benson, A. M., M. D.: E. P. Murdock, M. D.; H. Gradle, M. D.: G. Frank Lydston, M. D.; Hon. Frank J. Crawford, Counsellor at Law, and the Professors of the Science Department of Northwestern University.

TRUSTEES.—E. H. Sargent, D. R. Dyche, H. S. Maynard, T. H. Patterson, Wilhelm Bodemann.

CHICAGO COLLEGE OF PHARMACY.

NOS. 465 AND 467 STATE STREET.

Organized and incorporated in 1859.

Officers for 1888-89.—President, George Buck; Vice-President, Wm. K. Forsyth; Auditor, A. Emil Hiss; Treasurer, Judson S. Jacobus; Secretary, D. H. Galloway. Faculty.—Professor of Chemistry and Director of the Chemical

Laboratory.—H. D. Garrison. Professor of Pharmacy.--N. Gray Bartlett. Professor of Botany and Materia Medica and Director of Microscopical Laboratory.--E. S. Bastin. Professor of Pharmacal and Chemical Technology and Director of the Pharmaceutical Laboratories.--E. B. Stuart. Assistant Professor of Chemistry.—D. H. Galloway. Assistant Professor of Pharmacy.—L. C. Hogan. Assistant Professor of Materia Medica and Botany.- Wm. A. Puchner. Assistant in Prescription Department.—Richard A. Voge.

THE CHICAGO COLLEGE OF OPHTHALMOLOGY AND OTOLOGY.

Located at 527 State street, corner Harrison street. Is in a central part of the city and easy of access. It is designed for instructing physicians and those who wish to gain a higher knowledge in the science of Ophthalmology and Otology.

A free clinic is connected with it, that students may have practice in the art of applying the medicines to patients, which enables them to acquire practical knowledge while in the College. Tables will be set apart for students to work at during the time of the clinics. A large dark room is provided for students to practice in the use of the opthalmoscope, laryngoscope and rhinoscope, thus giving them the practice which they so much need. Lectures on refraction and optics will form a part of the instructions given during the course.

The importance of maintaining an Eye and Ear College cannot be too highly estimated, as physicians throughout the country are (as a rule) not qualified to examine and diagnose intra-ocular diseases, and he who is thus qualified has a very strong point in his favor.

Board of Trustees.—Henry Olin, M. D., Pres.; J. B. McFatrich, M. D., Vice-Pres.; Eli Wight, M. D.; A. L. Willard, M. D.; A. E. Bassett, M. D.; C. E. Starrett, M. D.

Faculty.—Henry Olin, M. D., Dean of the Faculty, 163 State street, Professor of Diseases of the Eye and Clinical Opthalmology; J. B. McFatrich, M. S., M D., Sec. of the Faculty, N. E. cor. State and Monroe Sts., Professor of Diseases of the Ear, Nose, and Throat and Clinical Otology; J. W. Tucker, M. D., 511 State street, Professor of Special Anatomy; A. L. Willard, M. D., 218 State street, Professor of Special Pathology: C. E. Starrett, M. S., M. D., 527 State street, Professor of Special Opthalmic and Aural Therapeutics, and Professor of Special Physiology: *Professor of Refraction, Accommodation and Optics; H E. Hilderbrand, M. D., 426 State street, Professor of Special Opthalmic and Aural Microscopy; E. Huntsinger, Room 39, 82 Madison street, Professor of Special Electro Therapeutics.

CHICAGO OPHTHALMIC COLLEGE,

607 WEST VAN BUREN STREET.

A clinical school of Ophthalmology, Otology and Laryngology, for graduates in medicine.

Faculty.—President, Professor of Diseases of the Eye and Clinical Ophthalmology, John E. Harper, A.M., M.D., 163 State street; Vice-President, Professor of Diseases of the Chest and Physical Diagnosis, S. A. McWilliams, A.M., M.D., 58 State street; Secretary, Professor of Diseases of the Ear and Clinical Otology,

*To be filled.

J. Brown Loring, M.D., M.R.C.S.Eng., 678 W. Lake street; Professor of Diseases of the Nose and Throat and Clinical Rhinology and Laryngology, Geo. F. Hawley, M.D., 125 State street; Professor of Diseases of the Nervous System, Oscar A. King, M.D., 70 Monroe street; Professor of Therapentics, C. C. P. Silva, M.D., 98 Ogden avenue; Professor of Clinical Laryngology and Rhinology, F. E. Waxham, M.D., 70 Monroe street; Professor of Histology and Pathological Anatomy, A. J. C. Saunier, M.D.; Professor of Clinical Ophthalmology, H. M. Martin, M.D.

The college has also established a department for Opticians.

WOMAN'S SCHOOL OF OPTICS AND OPHTHALMOLOGY.

A post graduate course for women. Rooms 208 and 209, 70 State St., Chicago Dr. Fannie Dickinson, President.

CHICAGO SCHOOL OF DERMATOLOGY.

Incorporated under the laws of the State of Illinois, 1887. For special clinical nstruction to graduates of medicine, in skin, genito-urinary, renal and rectal diseases, 605 W. Van Buren street, Chicago.

Faculty.—Professor of Dermatology and Genito-Urinary Surgery, Henry J. Reynolds President; Professor of Renal and Venereal Diseases, Arthur R. Reynolds, Secretary; Professor of Anatomy and Diseases of the Rectum, Chas. V. Bogue; Professor of Therapeutics, C. C. P. Silva; Professor of Physiology, Histology and Microscopy, John A Benson.

The course begins the first Tuesday in each month, and continues daily, except Sundays, for four weeks.

THE CHICAGO POLICLINIC.

A clinical school for practitioners of medicine. Corner Chicago and LaSalle avenues.

Officers 1888-89.—Truman W. Miller, M. D., Pres.; Daniel J. Avery, Esq., Vice-Pres., William T. Belfield, M. D., Sec.; John H. Chew, M. D., Treas.; A. E. Hoadley, M. D.

SOCIETIES.

INTERNATIONAL MEDICAL CONGRESS.

The tenth session of the International Medical Congress will be held in the city of Berlin in the year 1890. The officers are not elected until the assembling of the Congress, hence cannot be named at this time.

THE BRITISH MEDICAL ASSOCIATION

will hold its annual session in the city of Glasgow, Scotland, on the 8th day of August of the present year. Drs. N. S. Davis, S. J. Jones, A. E. Hoadley, D. A. K. Steele and F. E. Waxham are the Chicago delegates from the American Medical Association, and will represent that body at the coming session.

AMERICAN MEDICAL ASSOCIATION.

Organized in New York May 5th, 1846 nnder name of National Medical Convention. Reorganized and present name adopted in Baltimore in 1848.

Officers for 1888-89.—President, Dr. W. W. Dawson, Cincinnati, Ohio; 1st Vice-President, Dr. W. L. Schenck, Topeka, Kans.; 2d Vice-President, Dr. Frank Woodbury, Philadelphia; 3d Vice-President, Dr. H. O. Walker, Detroit, Mich.; 4th Vice-President, Dr. J. W. Bailey, Atlanta, Ga.; Permanent Secretary, Wm. B. Atkinson, 1400 Pine street, Philadelphia, Pa.; Treasurer, Richard J. Dunglison, lock box 1274, Philadelphia, Pa.; Librarian, C. H. A. Kleinschmidt, Washington, D. C. Next annual meeting Newport May 1889.

AMERICAN INSTITUTE OF HOMŒOPATHY.

Meets annually.

Officers,—Pres., Prof. A. C. Cowperthwait, M.D., Iowa City, Ia.; Vice-Pres., Prof. N. Schneider, M.D., Cleveland, Ohio; Treas., E. M. Kellogg, M.D., New York, N. Y.; General Secretary, Prof. P. Dudley, M.D., Philadelphia, Pa.; Provisional Secretary, T. W. Strong, M.D., Ward's Island, N. Y.

Annual meeting, June 25, 1888, at Niagara Falls. Forty-first session and forty-fifth anniversary.

NATIONAL ECLECTIC MEDICAL ASSOCIATION.

Organized in Cincinnati, May 25, 1848. Continued in active operations till 1857. Reorganized at Chicago, September 27, 1870.

Board of Officers 1888-9.—President, Milton Jay, M.D., Chicago, Ill.; Vice-Presidents, V. A. Baker, M.D., ——— Mich.; ——— Magrath, M.D., Atlanta, Ga.; ——— Montgomery, M.D., ——— Miss.; Secretary, Alexander Wilder, M.D., Newark, N. J.; Treasurer, James Anton, M.D., Lebanon, Ohio.

The nineteenth annual meeting for the election of officers will be held at Nashville, Tenn., June 18th, 1889.

AMERICAN SURGICAL ASSOCIATION.

Organized 1879. Meets annually the week preceding the meeting of the Amer. Med. Ass'n., in Washington. Its founder was Professor Samuel D. Gross. Membership limited to 100. Object: The Science and Art of Surgery. It publishes a yearly volume of transactions.

Officers 1888-89.—Pres., D. Hayes Agnew, M. D., Philadelphia, Pa.; Vice-Prest's., N. Senn, M. D., Milwaukee, Wis., F. S. Dennis, M.D., New York City; Recorder, J. Ewing Mears, M. D., Philadelphia, Pa.; Council, John S. Billings, M. D., U.S.A., Washington, D. C., L. McLane Tiffany, M.D., Baltimore, Md., R. A. Kinloch, M. D., Charleston, S. C.; Chairman of the Committee of Arrangements, John S. Billings, M.D., U.S.A., Washington, D. C.; Treas., P. S. Conner, M.D., Cincinnati, O.; Sec., J. R. Weist, M.D., Richmond, Ind.

AMERICAN PUBLIC HEALTH ASSOCIATION.

Organized April, 1872. "The members are elected with special reference to their acknowledged interest in or devotion to sanitary studies and allied sciences, and to the practical application of the same." Meets annually.

Organization 1888-89.- President, Dr. Charles N. Hewitt, Red Wing, Minn.; First Vice-President, Dr. G. B. Thornton, Memphis, Tenn.; Second Vice-President, Dr. Joseph Holt, New Orleans, La.; Secretary, Dr. Irving A. Watson, Concord, N. H.; Treasurer, Dr. J. Berrien, Lindsley, Nashville, Tenn. (*Ex-officio* Members Executive Committee.)

Executive Committee.- (Elective.)- Prof. George H. Rohe, Baltimore, Md.; Hon. D. P. Hadden, Memphis, Tenn.; Dr. Frederick Montizambert, Quebec, Canada; Dr. Henry B. Baker, Lansing, Mich.; Dr. S. H. Durgin, Boston, Mass.; Dr. J. N. McCormack, Bowling Green, Ky. The ex-presidents, *ex-officio* members Executive Committee. Dr. Stephen Smith, New York City; Dr. Joseph M. Toner, Washington, D. C.; Dr. Edwin M. Snow, Providence, R. I.; Dr. John H. Rauch, Springfield, Ill.; Prof. James L. Cabell, University of Virginia, Va.; Dr. John S. Billings, U. S. Army; Prof. Robert C. Kedzie, Lansing, Mich.; Dr. Ezra M. Hunt, Trenton, N. J.; Dr. Albert L. Gihon, U. S. Navy; Dr. James E. Reeves, Chattanooga, Tenn.; Dr. Henry P. Walcott, Cambridge, Mass.; Dr. George M. Sternberg, U. S. Army.

AMERICAN GYNÆCOLOGICAL SOCIETY.

Organized June 3, 1876, in New York, "for the promotion of knowledge in all that relates to the Diseases of Women and Obstetrics." Meets annually at such places as are determined by vote. President, Robert Battey, Rome, Ga.; Vice-Presidents, A. Reeves Jackson, Chicago, James R. Chadwick, Boston; Secretary, Joseph Talbot Johnson, Washington, D. C.; Treasurer, M. D. Mann, Buffalo. Other members of the Council: James B. Hunter, New York; R. B. Murry, Memphis; C. D. Palmer, Cincinnati; F. P. Foster, New York.

AMERICAN LARNYGOLOGICAL ASSOCIATION.

Organized in Buffalo, June 3, 1878, "for the promotion of knowledge in all that pertains to Diseases of the Upper Air Passages."

Officers 1888-89.—Pres., Rufus P. Lincoln, M. D., New York; 1st Vice-Pres., John N. MacKenzie, M. D., Baltimore; 2d Vice-Pres., Samuel W. Langmaid, M. D., Boston; Sec. and Treas., D. Bryson Delavan, M. D., 1 East 33d St., New York; Librarian, Thomas R. French, M. D., Brooklyn; Council, Frank Donaldson, M. D., Baltimore; J. Solis Cohen, M. D., Philadelphia; Franklin H. Hooper, M. D., Boston; E. Carroll Morgan, M. D., Washington.

AMERICAN OPHTHALMOLOGICAL SOCIETY.

Organized June 7, 1864, "for the advancement of Ophthalmic Science and Art." meets annually on the fourth Thursday in July, at such place as may be appointed by vote.

Officers,—President, Wm. F. Norris, M.D., Philadelphia; Vice-President, Hasket Derby, M.D., Boston; Secretary and Treasurer, O. F. Wadsworth, M.D., Boston; Corresponding Secretary, J. S. Prout, M.D., Brooklyn.

AMERICAN OTOLOGICAL SOCIETY.

Organized July 22, 1868.

Officers—President, Dr. J. S. Prout, Brooklyn, N. Y.; Vice-President, Dr. George C. Harlan, Philadelphia, Pa.; Secretary and Treasurer, Dr. J. J. B. Vermyne, New Bedford, Mass.; Committee on Membership, Drs. Gorham Bacon, D. B. St. John Roosa, J. Orne Green; Committee on Publications, Drs. J. J. B. Vermyne, C. J. Blake and J. Orne Green; Member of Committee on Organization of Congress of Special Societies, Dr. C. R. Agnew; alternate, Dr. W. H. Carmalt.

AMERICAN NEUROLOGICAL ASSOCIATION.

Officers.—President, J. J. Putnam, M.D., of Boston; Vice-President, Wharton Sinkler, M.D., of Philadelphia, and B. Sachs, M.D., of New York; Secretary and Treasurer, Graeme M. Hammond, M.D., 58 West 45th St., New York; Councillors, George W. Jacoby, M.D., of New York, and Robert T. Edes, M.D., of Washington.

AMERICAN DERMATOLOGICAL ASSOCIATION.

Officers.—President, I. E. Atkinson, M.D., Baltimore, Md.; Vice-President, P. A. Morrow, M.D., New York, N. Y.: Secretary and Treasurer, G. A. Tilden, Boston, Mass.

AMERICAN ASSOCIATION FOR THE CURE OF INEBRIATES.

Organized in 1870. Meets semi-annually.

Consists of superintendants, physicians and officers of inebriate asylums, and others interested in the study and treatment of inebriety as a disease. Publishes the *Quarterly Journal of Inebriety*, edited by the Secretary.

Officers.—Pres., Joseph Parrish, Burlington, N. J.; Vice-Pres., Albert Day, Boston, Mass.; 2d Vice-Pres., Louis D. Mason, Brooklyn, N. Y.; Sec. and Treas., T. D. Crothers, Hartford Conn.; Executive Com., L. D. Mason, J. E. Turner, C. H. Sheppard, and E. D. Mann.

THE AMERICAN CLIMATOLOGICAL SOCIETY.

Organized in New York, September 25, 1883, for the study of the influence of climate on the respiratory organs. Meets annually.

Officers.—President, Dr. A. L. Loomis, New York; Vice-Presidents, Dr. A. Y. P Garnett, Washington, Dr. J. T. Whittaker, Cincinnati; Secretary and Treasurer, Dr. J. B. Walker, Philadelphia, 1617 Green street. Council.—Dr. E. T. Bruen, Philadelphia; Dr. F. H. Bosworth, New York; Dr. F. C. Shattuck, Boston: Dr. R. G. Curtin, Philadelphia; Dr. A. H. Smith, New York. Next meeting in Washington in connection with the Congress of special societies, September 18, 19 and 20, 1888.

AMERICAN SOCIETY OF MICROSCOPISTS.

Organized 1878, for the promotion of Microscopical Science.

Officers,—President, David S. Kellicott, Ph.D., Buffalo, N. Y.: Vice-Presidents, H. J. Detmers, M. V. D., Columbus, Ohio, and T. B. Stowell, Ph.D., Cortland, N. Y.; Secretary, T. J. Burrill, Ph.D., Champaign, Ill.; Treasurer, S. M. Mosgrove, M.D., Urbana, Ohio. Members of the Executive Committee elected.—R. J. Nunn, M.D., Savannah, Ga.; C. C. Mellor, Pittsburgh, Pa.; H. D. Kendall, M.D., Grand Rapids, Mich. Ex-officio members.—R. H. Ward, M.D., Troy, N. Y.; H. L. Smith, LL. D., Geneva, N. Y ; J. D. Hyatt, Morrisania, N.Y.; Geo. E. Blackham, M.D., Dunkirk, N. Y.; Albert McCalla, Ph.D., Lake View, Ill.: Jacob D. Cox, LL.D., Cincinnati, Ohio; T. J. Burrill, Ph.D., Champaign, Ill.; W. A. Rogers, Waterville, Maine.

AMERICAN DENTAL ASSOCIATION.

Organized August 1869, at Saratoga. "The object of this Association shall be to cultivate the science and art of Dentistry, and all its collateral branches; to elevate and sustain the professional character of Dentists, to promote among them mutual improvement, social intercourse and good feeling, and collectively to represent and have cognizance of the common interests of the Dental Profession."

Officers, 1888-89.—Pres., W. K. Barrett, Buffalo, N. Y.; Vice-Pres., S. C. Ingersoll, Keokuk, Iowa; Rec. Sec., G. H. Cushing, Chicago; Cor. Sec., A. W. Harlan, Chicago; Treas., G. W. Keely, Oxford.

AMERICAN PHARMACEUTICAL ASSOCIATION.

ORGANIZED OCTOBER 7, 1852.

Officers of the Association.—President, John U. Lloyd, Cincinnati, O.; First Vice-President, Maurice W. Alexander, St. Louis, Mo.; Second Vice-President, Alexander K. Finlay, New Orleans, La.; Third Vice-President, Karl Simmon, St. Paul, Minn ; Treasurer, Samuel A. D. Sheppard, Boston, Mass.; Permanent Secretary, John M. Maisch, Philadelphia, Pa.; Local Secretary, James Vernor, Detroit, Mich.; Reporter on Progress of Pharmacy, C. Lewis Diehl, Louisville, Ky

The next meeting will be held at Detroit, Mich., September 3 to 7, 1888.

Standing Committees. Committee on Commercial Interests.— Chairman, Albert H. Hollister, Madison, Wis.; Secretary, Jos. W. Colcord, Lynn, Mass.; Edward A. Sayre, Newark, N. J.; William H. Rogers, Middletown, N. Y.; Alexander K. Finlay, New Orleans, La. Committee on Scientific Papers.—Chairman, T. Roberts Baker, Richmond, Va.; Secretary, A. B. Lyons, Detroit, Mich.; James M. Good, St. Louis, Mo. Committee on Prize Essays (Appointed by the Chairman of the section on Scientific Papers).—Chairman, George F. H. Markoe, Boston, Mass.; William W. Bartlet, Boston, Mass.; Leo Eliel, South Bend, Ind. Committee on Pharmaceutical Education.—Chairman, John F. Judge, Cincinnati, O.; Secretary, H. M. Whelpley, St. Louis, Mo.; P. W. Bedford, New York, N. Y. —Committee on Pharmaceutical Legislation.— Chairman, Randolph F. Bryant, Lincoln, Kan.; Secretary, W. P. DeForest, Brooklyn, N. Y.; John M. Maisch, Philadelphia, Pa. Committee on the Revision of the U. S. Pharmacopœia (Appointed by the President of the Association).Chairman, Albert E. Ebert, Chicago, Ill.; Albert B. Lyons, Detroit, Mich.; Alfred B. Huested, Albany, N. Y.; Alexander K. Finlay, New Orleans, La.; John H. Dawson, San Francisco, Cal.

Special Committees of the Association. Committee on Arrangements.— Chairman, James Vernor, Detroit Mich.; Theodore Ronnefeld, Detroit, Mich.; David O. Haynes, Detroit, Mich.; Frank Inglis, Detroit, Mich.; Thomas J. MacMahan, New York. Committee on Management.—Chairman, Joseph P. Remington, Philadelphia, Pa ; William S. Thompson, Washington, D. C.; Maurice W. Alexander, St. Louis, Mo.; Thomas J. Macmahan, New York, N. Y.; Samuel A. D. Sheppard, Boston, Mass.

MISSISSIPPI VALLEY MEDICAL SOCIETY.

Officers — President, Dudley Reynolds, Louisville, Ky.; Vice-President, A. Dunlap, Ohio; Vice-President, Dan. A. Thompson, Indiana; Vice-President, J. C. Fairbrother, Illinois; Vice-President, A. R. Jenkins, Kentucky; Vice-President, Y. H. Bond, Missouri; Treasurer, A. H. Ohmann-Dumesnill, St. Louis, Mo.; Permanent Secretary, J. L. Gray, Chicago, Ill.

Next annual meeting to be held at St. Louis, second Tuesday in September, 1888.

ILLINOIS STATE MEDICAL SOCIETY.

Organized 1850. Meets annually.

Officers 1888-89.—President, Chas. W. Earle, Chicago; 1st Vice-President, Philip H. Oyler, Mt. Pulaski; Secretary, D. W. Graham, Chicago; Assistant Secretary, T. M. Cullimore, Jacksonville; Treasurer, Thos. M. McIllvaine, Peoria.

Next meeting third Tuesday in May, 1889, at Jacksonville.

ILLINOIS STATE HOMŒOPATHIC MEDICAL ASSOCIATION.

Officers 1888-89.—Pres., Dr. Charles Gatchell, of Chicago; 1st Vice-Pres., Dr. F. W. Gordon, of Sterling; 2d Vice-Pres., Dr. W. A. Smith, of Winona; 3d Vice-Pres., Mrs. Dr. E. H. Stansberry, of Chicago; Treas., Dr. A. A. Whipple, of Quincy; Secretary, Dr. A. B. Spaech, of Englewood. Board of Censors—Dr. L. Pratt of Wheaton, Dr. J. W. Coyner of Peoria, Dr. W. J. Hawkes of Chicago, Dr. C. A. Weirick of Marseilles, and Dr. J. S. Mitchell of Chicago.

Chairmen of Committees or Bureaus for the coming year.—Materia Medica, Dr. F. W. Gordon, of Sterling; Medical Legislation, Jurisprudence, and Education, Dr. J. A. Vincent, of Springfield, Ill.; Ophthalmology and Otology, Dr. C. H. Vilas, of Chicago; Clinical Medicine, Dr. J. B. Dunham, of Winona: Surgery, Dr. W. F. Knoll, of Chicago; Diseases of Children, Dr. M. J. Hill, of Sterling; Sanitary Science and Hygiene, Dr. C. W. Harback, of Lockport: Pharmacy, Dr. James E. Gross, of Chicago; Medical Literature, Dr. Charles Gatchell, of Chicago: Necrology, Dr. T. S. Hoyne, of Chicago.

ILLINOIS STATE ECLECTIC MEDICAL SOCIETY.

Organized in 1869. Officers for 1888-89.—President, R. F. Bennett, M. D., Litchfield; 1st Vice-President, F. P. Antle, M. D., Petersburg: 2d Vice-President, C. V. Massey, M. D., Athens; Rec. Secretary, W. E. Kinnett, M. D., Yorkville: Cor. Secretary, G. R. Shaffer, M. D., Morton; Treasurer, C. A. Doss, M. D., Pittsfield. Next meeting at Springfield, May 15 and 16, 1889.

ILLINOIS STATE DENTAL SOCIETY.

President, Geo. H. Cushing, Chicago; Secretary, Garrett Newkirk, Chicago.

ILLINOIS PHARMACEUTICAL ASSOCIATION.

Organized 1880. Meets annually. The object of this organization shall be to promote the interests of pharmacy, by urging the enactment of such laws as will be of mutual advantage to pharmacists and the public, by restricting the dispensing and sale of medicines to competent parties, to encourage a more thorough training of assistants, and, finally, to bring the pharmacists of this State into more intimate social relations.

Officers—President, Henry Smith, Decatur; Vice-Presidents, W. P. Boyd, Arcola, J. E. Espey, Chicago, C. F. Prickett, Carbondale; Secretary, L. C. Hogan, Englewood; Treasurer, C. A. Strathman, El Paso; Executive Committee, C. H. Plautz, Chicago, G. H. Sohrbeck, Moline, T. C. Loehr, Carlinville; Editor, L. C. Hogan, Englewood; Local Secretary, W. M. Benton, Peoria.

Next meeting at Peoria, August 16, 1888.

THE CHICAGO MEDICAL SOCIETY.

Organized April 5, 1852, as the Cook County Medical Society. In 1858 the name was changed to the Chicago Medical Society. Meetings, the first and third Mondays of every month, at 8 p. m. The annual meeting, at which election of officers and delegates occurs, is the first Monday in April. "Any physician of good standing in the profession shall be eligible for membership."

List of Officers and Standing Committees for the year 1888-89.—Dr. J. H. Etheridge, President; Dr. A. E. Hoadley, First Vice-President; Dr. R. D. MacArthur, Second Vice-President; Dr. Frank Billings, Secretary: Dr. Frank S. Johnson, Treasurer; Dr. J. H. Wilson, Necrologist.

Committee on Judiciary.—Dr. Wm. E. Quine, term expires April, 1891; Dr. E. J. Doering, term expires April, 1890; Dr. A. H. Foster, term expires April, 1889.

Committee on Membership.—Dr. L. H. Montgomery, term expires April, 1891; Dr. F. E. Waxham, term expires April, 1890; Dr. G. C. Paoli, term expires April, 1889.

. Committee on Library.—Dr. J. Bartlett, term expires April, 1891; Dr. D. W. Graham, term expires April 1890; Dr. E. Andrews, term expires April, 1889.

Committee on Publication.—Dr W. T. Thackeray, term expires April, 1891; Dr. S. H. Stevenson, term expires April, 1890; Dr. J. A. Robison, term expires April, 1889.

CHICAGO ACADEMY OF HOMŒOPATHIC PHYSICIANS AND SURGEONS.

Organized in 1868. The "Object shall be the advancement and improvement of Homœopathy, and the collateral branches of medical science."

Officers 1887–89.—Lemuel C. Grosvenor, Pres.; F. H. Gardiner, M.D., Vice-Pres.; R. W. Conant, Sec. and Treas.

CHICAGO ECLECTIC MEDICAL AND SURGICAL SOCIETY.

Organized 1874. Reorganized, 1878. Regular meetings at the Grand Pacific Hotel on the third Wednesday of each month. Annual meeting, third Wednesday in December.

The objects of this Society are: Association and co-operation for mutual benefit, the promotion of professional interests, the communication and development of useful and scientific knowledge, the cultivation of a *liberal* and fraternal sentiment and the advancement of the science and art of medicine. "With charity for all—with malice toward none."

Officers, 1888-89.—President. E. F. Buecking; Secretary, Oscar O. Baines; Treasurer, H. K. Stratford, 243 State Street.

WOMEN'S HOMŒOPATHIC MEDICAL SOCIETY OF CHICAGO.

Officers.—President, Lucy Wait, M. D.; Vice-President, Anna Parke; Secretary, I. L. Green.

CHICAGO MEDICO LEGAL SOCIETY.

For the study and advancement of the science of medical jurisprudence.

The Society meets quarterly on the first Saturday in June, September, December and March.

Officers 1888-89.—President, E. J. Doering, 2406 Prairie avenue; First Vice-President, B. Bettman, 18 and 19 Central Music Hall; Second Vice-President, Eric Winters; Treasurer, L. L. McArthur; Secretary, Scott Helm.

CHICAGO MEDICO-HISTORICAL SOCIETY.

Organized April 28, 1874. The objects of this society are, "To discover, procure and preserve whatever may relate to the medical history of Chicago and vicinity, and the publication of such information as may be, from time to time, determined upon. Its work so far has been confined chiefly to the publication of

The Annual Medical Register, which is conducted by an editor, with the assistance of a committee on publication."

The list of physicians for the Chicago Register is submitted annually to a vote of the whole society. Regular meetings are held on the last Tuesdays of April, July, October and January; that of April being the annual meeting.

The committee on publication nominate candidates for membership at any regular or special meeting, and the names are voted on at any subsequent meeting. All the officers are elected annually, except the editor, and the president and editor are ex-officio members of the committee on publication.

Officers for 1888-89.- President, Eugene Marguerat; Vice-President, J. Ramsey Flood; Secretary and Treasurer, P. S. Hayes; Diarist, Liston H. Montgomery; Editor, D. W. Graham.

CHICAGO PATHOLOGICAL SOCIETY.

Meetings second Monday in the month at Park Institute, 90 S. Ashland ave.

Officers 1888-89.—President, I. N. Danforth; Vice-President, J. D. Skeer; Secretary, W. L. Copeland; Treasurer, Jos ph Haven.

CHICAGO GYNÆCOLOGICAL SOCIETY.

Organized October 18, 1878. Chartered June 8, 1880.

The meetings are held on the third Friday evening of each month.

Officers, 1888-89.—President, Henry T. Byford; Vice-President, Addison H. Foster; Editor, W. W. Jaggard; Secretary and Treasurer, Ed. Warren Sawyer.

CLINICAL SOCIETY OF HAHNEMANN.

Organized February 6, 1877. Its object is "Medical co-operation in the highest and best sense of the term, to the end that each of its members may have the benefit of the observations, best thoughts, and practical results of its entire membership." The meetings are held the first Saturday in each month at the Grand Pacific Hotel. The annual meeting is in April. Eligible for membership: Those who are graduates, and in good standing, of a recognized college of medicine.

Officers 1888-89.—President, G. F. Shears; Vice-Presidents, C. E. Laning; J. D. Craig, E. Parkhurst; Secretary, Jos. P. Cobb, M.D.; Treasurer, W. M. Davison; Phonetic Reporter, J. B. S. King; Board of Censors, H. B. Fellows, T. S. Hoyne, W. H. Burt, W. S. Gee; Executive Committee, G. A. Hall, E. M. P. Ludlam, A. J. French, W. J. Hawkes.

CHICAGO SOCIETY OF OPHTHALMOLOGY AND OTOLOGY.

Organized September 21, 1883. Meetings are held on the second Tuesday of every month, at the Tremont House. Annual meeting will be held second Tuesday in October.

"Members must consist of graduates from reputable schools of medicine, who are engaged in the practice of ophthalmology or otology."

Officers.—President, E. L. Holmes, 112 Clark street; Vice-President, Lyman Ware, 125 State Street; Secretary and Treasurer, Boerne Bettman, 18 and 19 Central Music Hall.

CHICAGO DENTAL SOCIETY.

Organized 1864. Reorganized 1878. Meets at Chicago College of Dental Surgery, first Tuesday evening of each month, except August and September.

Officers, 1888-89.—President, James A. Swasey; Secretary, C. N. Johnson.

ODONTOLOGICAL SOCIETY.

President, Edmond Noyes; Secretary, P. J. Kester.

CHICAGO DENTAL CLUB.

President, A. B. Freeman; Secretary, C. S. Smith.

ODONTOGRAPHIC SOCIETY.

President, C. E. Bentley; Secretary, Geo. N. West.

THE SCANDINAVIAN MEDICAL SOCIETY.

Was organized Oct. 21st, 1887. Its object is to promote friendly feeling among its members, to encourage professional zeal and to interchange professional experience. Any regular Scandinavian physician in good standing, in United States, is eligible for membership. The meetings are held the third Wednesday of each month, at Sherman House; the annual meeting in October.

Officers 1887-88.—President, S. D. Jacobson; Vice-President, F. A. Hess; Sec'y and Treas., Sven Windrow.

ALUMNI ASSOCIATION OF THE CHICAGO MEDICAL COLLEGE.

Officers and Committees for the year 1888-89.—President, F. C. Schaefer, Class of 1876, Chicago; 1st Vice-President, J. L. Sawyers, Class of 1878, Centerville, Ia.; 2d Vice-President, W. D. Storer, Class of 1888, Chicago; Secretary, Horace M. Starkey, Class of 1878, Chicago; Treasurer, E. W. Andrews, Class of 1881, Chicago; Necrologist, G. W. Webster, Class of 1882, Chicago.

Committee on the Correlation of the College and Alumni.—Chairman, Horace M. Starkey, Chicago; E. J. Doering, Chicago; H. T. Byford, Chicago; Frank Billings, Chicago; George W. Jones, Danville, Ill.

Committee on Permanent Membership Fund.—Chairman, A. H. Burr, Chicago; Elbert Wing, Chicago; J. L. Gray, Chicago.

Committee on Library.—Chairman, M. P. Hatfield, Chicago; E. W. Andrews, Chicago; N. S. Davis, Jr., Chicago.

. Committee on Medical Jurisprudence.—Chairman, D. A. Sheffield, Apple River, Ill.; M. D. Ewell, Chicago; F. C. Schaefer, Chicago; E. W. Andrews, Chicago; H. T. Byford, Chicago.

Committee on Preliminary Education.—Chairman, Professor J. A. Davies, Madison, Wis.; N. S. Davis, Jr., Chicago, Ill.; C. W. Earle, Chicago, Ill.; L. T. Potter, Chattanooga, Tenn.; E. J. Doering, Chicago.

Committee on Fowler Prize.—Chairman, Horace M. Starkey, Chicago; Lyman Ware, Chicago; F. C. Schaefer, Chicago.

ALUMNI ASSOCIATION OF RUSH MEDICAL COLLEGE.

Officers for the year 1888-9.—President, L. D. Dunn, Moline, Ill.; First Vice-President, J. R. Washburn, Renssler, Ind.; Second Vice-President, J. J. M. Angear, Chicago; Secretary and Treasurer, F. A. Emmons, M.D., Chicago; Executive Committee, E. B. Weston, M.D., Chicago; D. B. Conley, M.D., Chicago; S. Cole, M.D., Chicago; Auditing Committee, A. H. Wimermark, M.D., Chicago; F. A. Hess, Chicago.

ALUMNI ASSOCIATION OF THE COLLEGE OF PHYSICIANS AND SURGEONS OF CHICAGO.

Officers for 1888-9.—President, Dr. E. E. Babcock; First Vice-President, Dr. J. R. Williams; Second Vice-President, Dr. G. S. Harkness; Third Vice-President, Dr. W. F. Malone; Fourth Vice-President, Dr. C. B. Wood; Corresponding Secretary, Dr. A. L. Wagner; Recording Secretary, Dr. T. A. Davis; Treasurer, Dr. R. W. Jones; Historian and Librarian, Dr. L. Hektoen; Executive and Publishing Committee, B. L. Merril, C. A. Earle, D. B. Wiley.

WOMAN'S MEDICAL COLLEGE, ALUMNÆ ASSOCIATION.

OFFICERS FOR THE YEAR 1888—9.

President, Mary E. Bates, Chicago, class 1881; Secretary, Eliza H. Root, Chicago, Ill., class of 1882; Treasurer, Elizabeth Holton, Englewood, Ill.; Necrologist, Emma Nichols Wanty, Grand Rapids, Mich.; Auditors.—Fanny Dickinson, Chicago, Ill., Catherine B. Slater, Aurora, Ill.; Committee on Membership, Mary E. Bates, Chicago, Ill., Isabel R. Copp, North Port, Mich., Lucind Corr, Collinville, Ill.; Executive Committee.—Mary E. Bates, Pres., Eliza H. Root, Sec., Elizabeth Holton, Treas., Jennie E. Hayner, Chicago, Ill., Marie J. Mergler, Chicago, Ill., Abbie Mare, Chicago, Ill.

HAHNEMANN COLLEGE ALUMNI ASSOCIATION.

Officers.—President, Dr. M. H. Parmelee; Vice-President, Dr. Wm. Davidson; 2d Vice-President, Dr. J. H. Thompson; Secretary, Dr. H. V. Halbert; Treasurer, Dr. J. P. Cobb; Necrologist, Dr. M. H. Landuth.

CHICAGO HOMŒOPATHIC MEDICAL COLLEGE ALUMNI ASSOCIATION.

Officers, 1888-89.—President, W. J. Bliem; Vice-President, C. C. Dodge; Secretary, W. M. Stearns; Treasurer, S. N. Schneider.

BENNETT COLLEGE ALUMNI ASSOCIATION.

(List of officers not received in time for insertion.)

HOSPITALS.

MERCY HOSPITAL.

Telephone 8267.

CORNER OF CALUMET AVENUE AND TWENTY-SIXTH STREET.

Conducted by the Sisters of Mercy. Founded about 1848–49, by the Sisters of Mercy, and was incorporated in 1852 as "Mercy Hospital and Mercy Orphan Asylum."

Medical Staff.—Physicians, N. S. Davis, Sr., N. S. Davis, Jr., J. H. Hollister; Consulting, H. A. Johnson and E. O. F. Roler; Surgeons, Edmund Andrews and E. Wyllys Andrews; Gynaecologists, E. C. Dudley and F. T. Andrews; Ophthalmologist and Aurist, H. M. Starkey; Oral Surgeon, J. S. Marshall; Obstetrician, W. W. Jaggard: House Staff, M. Scheuer, B. L. Riese and W. D. Storer.

HAHNEMANN HOSPITAL,

GROVELAND AVENUE, REAR OF COLLEGE BUILDING.

Telephone 8104.

Officers.—President, D. S. Smith; Dean, Reuben Ludlam; Treasurer and Bus. Manager, T. S. Hoyne; Register, E. S. Barley; Superintendent, G. F. Shears.

Medical Staff.—Diseases of Women, R. Ludlam, M.D.; Surgery, Prof. G. A. Hall, M.D.; Surgery, G. F. Shears, M.D.; Nervous Diseases, H. B. Fellows: Skin and Venerial Diseases, T. S. Hoyne: Diseases of Women, E. S. Bailey: Eye and Ear, C. H. Vilas; Eye and Ear, Dr. Joseph Watry; General Medicine, W. J. Hawkes; Physical Diagnosis, B. J. Arnulphy; General Medicine, Charles E. Laning; Obstetrics, S. Leavitt; House Surgery, H. R. Chislett; House Physician, V. W. Stiles.

ST. LUKE'S FREE HOSPITAL.

NO. 1434 INDIANA AVE.

(Tel. 8138.) This noted institution of the city originated with a few benevolent ladies during the civil war, in an effort to care for the sick prisoners and soldiers at Camp Douglass. In 1864 it was started as a free hospital for the sick poor, and has reached its present eminence and prosperity after many trials and removals.

It is under Episcopalian control and management, but no distinction is made in admission on account of sect, sex or nationality.

Medical Board.—Attending Surgeons, John E. Owens, M.D., L. L. McArthur, M.D.; Attending Physicians, I. N. Danforth, M.D., James H. Etheridge, M.D.; Attending Obstetricians, G. M. Chamberlin, M.D., De Laskie Miller, M.D.; Gynecologists, E. C. Dudley, H. T. Byford; Attending Oculists and Aurists, Samuel J. Jones, M.D., and R. Tilley M. D.; Pathologists, Frank Johnson, M. D. Frank Cary, M.D.; Dental Surgeons, W. W. Allport, M.D., D.D.S., and John S. Marshall M. D.

Dispensary Staff.—Attending Physician, Dr. L. B. Hayman; Attending Surgeons, Robert Tilley, M.D.; S. J. Jones, M.D.; Attending Gynæcologists, Frank Cary, M.D., and J. C. Hoag, M. D.; Attending Surgeons, H. H. Frothingham; Lewis L. McArthur, M.D.; Internes, T. B. Schwartz, M.D.,——Schwant, M.D., L. L. Gregory, M.D.

THE CHICAGO HOSPITAL FOR WOMEN AND CHILDREN.

WEST ADAMS, CORNER OF PAULINA ST.

Founded in 1865. A commodious building has been recently erected. There are now fifty ward beds.

The following staff are in attendance:—Gynæcology and Gynæcological Surgery, Prof. Mary H. Thompson; General Diseases, Prof. Sarah H. Stevenson; Obstetrics, Prof. Eliza H. Root; Surgery, Dr. Mary A. Mixer; Diseases of Children, Dr. Harriet Hoyl Cary; Pathology, Dr. Annette S. Dobbin.

ST. JOSEPH'S HOSPITAL.

360 GARFIELD AVE., CORNER BURLING ST., CHICAGO.

Telephone 3543.

St. Joseph's Hospital was established in 1869, by the Sisters of Charity. It is situated in the northern section of the city, within one-fourth mile of Lake Michigan and Lincoln Park, and is of easy access by four lines of street cars. The location is in the highest and dryest portion of the city, and from its proximity to the Lake and Park, is one of the healthiest and pleasantest in the west. Both males and females are received. Physicians whose names are not on the Staff, have the privilege of attending patients occupying private rooms. Patients are received for the treatment in General and Orthopædic Surgery, Nervous and Mental Diseases, the Diseases of Women, Lying in Cases, and in Medicine generally. Many poor patients are annually treated free of charge.

Medical Staff.—Surgeon in Charge, Prof. Charles T. Parkes, residence 51 Lincoln avenue; Physician in Charge, Prof. F. L. Wadsworth, M.D., residence 31 Dearborn avenue; Mental and Nervous Diseases, Prof. Dan R. Brower, M.D., residence 571 West Adams street; Diseases of Throat and Nose, Prof. E. F. Ingals, M.D., residence 637 West Adams street; Obstetrician and Diseases of Children and House Physician, Dr. Geo. W. Reynolds, residence 282 Bissell street; Diseases of Eye and Ear, Dr. Robert Tilley, residence Matteson House; Diseases of Genito Urinary Organs, Dr. E. W. Whitney, residence 289 West 12th street; House Surgeon, Dr. C. W. Johnson, residence 112 Chicago avenue; Consulting Surgeons, Prof. W. Godfrey Dyas, M.D., residence 107 Clark street, Prof. R. G. Bogue, residence 5 Washington Place; Consulting Physician, Prof. J. Adams Allen, M.D., residence Woodruff Hotel; Visiting Surgeon, Dr. C. F. Korsell.

Out Patient Department.—This provides for the treatment of all persons presenting themselves at the Hospital office, and includes the same wide range of treatment as the Hospital proper. Free to those without means.

WOMAN'S HOSPITAL.

Telephone 8353.

CORNER OF RHODES AVENUE AND THIRTY-SECOND STREET.

Was founded in 1870 as the "Woman's Hospital of the State of Illinois," by Dr. A. Reeves Jackson, especially devoted to the treatment of diseases and accidents peculiar to woman; the clinical instructions of students in medicine, and the practical training of nurses.

Officers.—President, Mrs. A. H. Barber; Vice-President, Mrs. L. H. Bisbee,. Secretary, Mrs. T. Burnham; Treasurer, Mrs. J. A. Perkins; Medical Staff, W. H Byford, H. P. Merriman, Daniel T. Nelson, A. Davenport Piercy, Marie J. Mergler; Henry T. Byford; Resident Physicians, Roda M. Galloway, Louisa Sedgwick; Consulting Physicians, H. A. Johnson, De Laskie Miller, Daniel R. Brower, Sarah Hackett Stevenson; Superintendent Training School, Frances Cantrell.

THE HOSPITAL OF THE ALEXIAN BROTHERS.

569 N. MARKET STREET,

Represents the first establishment of the order in America. After the fire of 1871 the spacious building now occupied was erected at a cost of $45,000. Only men are admitted, but there is a free dispensary in connection which is open to the sick and needy of both sexes. There is no discrimination as to creed or nationality. Many patients are sent in by the city, but the expenses, which amount to about $18,000 per annum, are met wholly by subscriptions solicited by two of the Brothers.

Medical Staff.—Consulting Physician, Ernst Schmidt; Attending Physicians, Rud. Seiffert, M. Mannheimer; Attending Surgeon, A. J. Baxter; Nervous and Mental Diseases, S. V. Clevenger; Ophthalmic Surgeon, F. C. Hotz; Pathologist, Chas. S. Bacon; Residence Physician, J. W. Oswald.

THE CENTRAL HOMŒOPATHIC HOSPITAL AND FREE DISPENSARY.

Telephone 7291.

Organized in 1876, and is located in the Homœopathic Medical College building, on the corner of Wood and York streets. It furnishes medicines and surgical assistance free to the poor, and attends obstetric cases gratis.

Hospital and Dispensary Staff.—Superintendent, C. M. Beebe, M.D.; Medical Department, J. S. Mitchell, J. R. Kippax, N. B. Delamater, A. W. Woodward, R. N. Tooker; Surgical Department, A. G. Beebe, E. H. Pratt, W. F. Knoll; Eye and Ear Department, J. H. Buffum, W. M. Stearns; Gynæcological Department, J. W. Streeter, C. M. Beebe: Obstetrical Department, R. N. Foster; Dental Department, F. H. Gardiner.

House Staff.—House Physician, D. A. Foote; 1st Assistant, C. Frazer, M.D.; 2d Assistant, Joseph Low, M.D.

The college clinics are held regularly throughout the year as follows: Monday, 11:30 A. M., Medical, Prof. J. S. Mitchell; Tuesday, 10:30 A. M., Surgical and Venereal, Prof. E. H. Pratt; Tuesday, 3 P. M., Gynæcological (sub-clinic), Prof. J. W. Streeter, C. M. Beebe, M.D.; Wednesday, 2 P. M., Eye and Ear, Prof. J. H.

Buffum, Dr. W. M. Stearns; Wednesday, 3 P. M., Neurological, Prof. N. B. Delamater, E. W. Keith, M.D.; Thursday, 11 A. M., Medical and Skin, Prof. J. R. Kippax, Prof. A. W. Woodward, Dr. M. J. Bliem; Friday, 3 P. M., Gynæcological, Prof. J. W. Streeter, C. M. Beebe, M.D.; Saturday, 10:30 A. M., Surgical, Prof. A. G. Beebe, Dr. W. F. Knoll; Saturday, 11:30 A. M., Children's, Prof. R. N. Tooker.

PRESBYTERIAN HOSPITAL.

(Telephone 7189.)

Corner Wood and West Harrison streets.

Officers 1888-89.—Daniel K. Pearson, Pres.; W. A. Douglas, Rec. Sec.; John S. Gould, Vice-Pres.; J. A. Robison, Asst. Rec. Sec ; Geo. W. Hale, Treas.

Medical Board.—Joseph P. Ross, M. D., President; J. A. Robison, M. D., Secretary. Consulting Physicians—J. Adams Allen, M. D., Charles Gilman Smith, M. D. Consulting Surgeons—R. N. Isham, M. D., Roswell G. Bogue, M. D. Attending Physicians—Joseph P. Ross, M. D., Norman Bridge, M. D., H. M. Lyman, M. D. Attending Surgeons—C. T. Parkes, M. D., D. W. Graham, M. D., E. W. Whitney, M. D. Attending Gynæcologists—James H. Etheridge, M. D., H. P. Merriman, M. D., Philip Adolphus, M. D. Attending Physicians for Diseases of Children and Accouchers—DeLaskie Miller, M. D., J. Suydam Knox, M. D. Attending Dermatologists—J. Nevins Hyde, M. D., R. D. MacArthur, M. D. Attending Oculists and Aurists—E. L. Holmes, M. D., Lyman Ware, M. D. Attending Physician for Throat Disease—John A. Robison, M. D. Medical Superintendent—H. B. Stehman, M. D. Curator—French Moore, M. D.

THE MICHAEL REESE HOSPITAL

Maintained and managed by the Hebrew Relief Association is located on the corner of Twenty-ninth street and Groveland avenue. It was erected from a fund provided by the will of the late Michael Reese. There is no test of faith in admission, and both male and female patients are received.

Medical Staff.—Ernst Schmidt, Michael Mannheimer, Henry Banga, Henry Gradle, James Nevins Hyde, L. L. McArthur, Frank S. Johnson. Consulting Staff.—Edmund Andrews, A. J. Baxter, S. V. Clevenger, H. A. Johnson. House Physician, Louis Greensfelder.

BENNETT HOSPITAL.

This institution is situated upon the College grounds, adjacent to the College building, and connected with it by covered ways.

Hospital Staff.—Milton Jay, Surgeon; J. B. McFatrich, Ophthalmologist; A. L. Clark, Gynæcologist; H. S. Tucker, Resident Physican.

AUGUSTANA HOSPITAL AND DEACONESS INSTITUTE,

Telephone 3022.

Of the Swedish Ev. Luth. Church, at 151 Lincoln Ave. (tel. 3022), was organized in 1882 as a charitable institntion. In May, 1884, hospital work was begun with the view of forming the nucleus of an establishment to meet the needs of the denomination of the Northwest.

Board of Trustees.—Rev. Erl. Carlsson, Andover, Ill.; Rev. M. C. Ranseen, Cor. Huron and May Sts., Chicago, Ill.; Rev. C. A. Evald, 218 Sedgwick St., Chicago, Ill.; Rev. L. G. Abrahamson, 2815 Portland Ave., Chicago, Ill.; G. A. Bohman, 175 E. Chicago Ave., Chicago, Ill.; John Erlander, Rockford, Ill.; C. W. Smith, Treasurer, 117 Chicago Ave., Chicago, Ill.

Medical Staff.—Surgeon-in-Chief, Truman W. Miller; Attending Physician, John H. Chew; Attending Physician, P. M. Woodworth; Matron, Lotta Fried.

THE MAURICE PORTER MEMORIAL HOSPITAL,

606 FULLERTON AVENUE, CORNER ORCHARD STREET.

Was founded in 1882, by Mrs. Julia Porter, in memory of her dead son. It is dedicated exclusively to the free care and treatment of children between three and thirteen years of age. Children suffering from incurable or contagious diseases cannot be admitted. Patients will be admitted on application to either of the surgeons or to the superintendent at the hospital, and will be received at all times without formality.

Medical Staff.—Surgeons, Truman W. Miller; William T. Belfield; Assistant Surgeon, F. D. Porter; Physicians, Charles L. Rutter; Frank Billings; Occulist and Aurist, George T. Fiske; Superintendent, Miss Headline.

NATIONAL TEMPERANCE HOSPITAL,

3411 COTTAGE GROVE AVE.

Telephone 9981.

Board of Trustees.—President, Mrs. J. B. Hobbs, 343 La Salle avenue, Chicago; Corresponding Secretary, Dr. Mary Weeks Burnett, 1 Central Music Hall, Chicago, Ill.; Recording Secretary, Mrs. Mary B. Willard; Treasurer, Mrs. L. H. Plumb, Streator, Ill.; Mrs. E. N. Peters, Manistee, Mich., Mrs. J. K. Barney, Providence, R. I., Miss Mary Allen West, 161 La Salle street, Chicago.

EMERGENCY HOSPITAL.

184 to 192 Superior Street, between La Salle Avenue and Wells Street. Supported by voluntary contributions. Receives both free and paying patients.

Attending Surgeons.—Ralph N. Isham, Christian Fenger; Orderly Surgeon, Albert Swan; Matron, Friedericka Berkne.

HOME FOR INCURABLES.

Incorporated May 26, 1880. Corner Fullerton and Racine Aves., was opened for the reception of patients, January, 1881. The building, a two-story brick, was formerly a private residence; it stands on a lot of two acres. The institution provides a home for all classes of incurable patients, except the insane and those suffering from the acute and infectious stages of contagious diseases, such home supplying medical and surgical aid to its inmates.

Medical Staff.—A. H. Cooke, A. P. Gilmore, H. C. Hutchinson, W. H. Byford, De Laskie Miller, Chas. T. Parkes, R. D. Mac Arthur, J. Mason Cooke, J. H. Wilson, J. King Winer, W. H. Weaver, Wm. Goodsmith; Matron, Mrs. C. S. Barlow.

THE CHICAGO SANITARIUM.

Located on Ashland avenue, has been instituted during the past year for the cure and treatment of mental and nervous diseases. Its central location in the heart of the city, renders it particularly desirable. The accommodations at the Sanitarium are everything that could be wished, the attendants are nurses educated especially in the care of diseases of the character to which the institution is devoted, and the Medical Staff is composed entirely of physicians who have made the treatment of nervous diseases a life study. J. L. Gray M.D., Physician in charge.

LAKESIDE HOSPITAL.

3545 Vincennes Avenue, for the treatment of rectal diseases, exclusively. It has been specially fitted up for that purpose, and has an efficient corps of nurses and attendants, under the immediate supervision of E. H. Dorland, M.D., Physician in charge. Superintendent, Edwin Purlier; Matron, Mrs. E. Purlier.

CHICAGO GYNÆCOLOGICAL INSTITUTE,

for the Treatment of the Medical and Surgical Diseases of Women, No. 5306 Jefferson avenue, Hyde Park. Under the management of Dr. Lucy Waite and Dr. Clara W. Peaslee.

SOUTH SIDE FREE DISPENSARY,

26TH ST. AND PRAIRIE AVE., CHICAGO MEDICAL COLLEGE BUILDING.

President, J. R. Kewley; Vice-President, W. E. Casselberry; Secretary, D. H. Williams.

Staff.—W. W. Haggard. W. E. Casselberry. E. W. Andrews, M. P. Hatfield, H. M. Starkey, W. E. Morgan, G. S. Isham, R. E. Starkweather, E. C. Helm, P. S. Hayes, D. H. Williams, F. M. Wilder, J. R. Kewley, F. H. Martin, R. H. Babcock. C. H. Lodor, Lester Curtis, Chas. Caldwell, J. H. Besharian, J. C. Cook, R. Randolph, J. L. Gray, A. R. Small, A. G. Paine and C. A. Foulks.

WEST SIDE FREE DISPENSARY,

315 Honore street, cor of Harrison, was established and incorporated September 9, 1881, as a clinical annex of the College of Physicians and Surgeons, and for the gratuitous treatment of the deserving poor. The dispensary is located on the first floor of the college.

Medical and Surgical Staff.—President and Chief of Medical Department, S. A. McWilliams; Chief of Surgical Department, H. Palmer; Chief of Gynæcological Department, A. Reeves Jackson; Chief of Children's Department, Henry P. Newman; Chief of Eye Department, John E. Harper: Chief of Dental Department, A. W. Harlan; Chief of Neurological Department, Oscar A. King; Chief of Laryngological Department, Frank E. Waxham: Chief of Dermatological Department, Henry J. Reynolds.

Attending Physicians and Surgeons.—In the Medical Department, Drs. A. R. Reynolds, J. M. Patton, Andrew Stewart, Ira Munzer, B. A. Brigham and R. W. Jones: in the Surgical Department, Drs. Thos. A. Davis and Weller Van Hook; in the Gynæcological Department, Drs. H. P. Newman, Frank B. Earle, W. C.

Caldwell and A. J. C. Saunier; in the Children's Department, Dr. S. S. Bishop; in the Eye and Ear Department, Drs. J. B. Loring and H. M. Martin; in the Department for Nervous Diseases, Dr. John Fisher; in the Department for Genito-Urinary Diseases, Drs. G. Frank Lydston and H. J. Reynolds; in the Department for Diseases of the Nose and Throat, Dr. Geo. A. Thomson; in the Department for Diseases of the Skin, Dr. Henry J. Reynolds.

CENTRAL FREE DISPENSARY.

CORNER WOOD AND HARRISON STREETS.

(Telephone 7147.)

Medical Staff. Consulting Physicians and Surgeons.—R. C. Bogue, M. D., W. H. Byford, M. D., E. L. Holmes, M. D., H. M. Lyman, M. D., J. P. Ross, M. D. Attending Physicians and Surgeons.—Philip Adolphus, M. D., Wm. T. Belfield, M. D., Norman Bridge, M. D., T. W. Brophy, M. D., S. N. Chapin, M. D., B. Dorr Colby, M. D., E. J. Colburn, M. D., A. C. Cotton, M. D., Jas. H. Etheridge, M. D., D. W. Graham, M. D., J. E. Heguembourg, M. D., Alfred Hinde, M.D., F. A. Hodson, M. D., J. W. Hutchins, M. D., E. Fletcher Ingals, M. D., A. E. Kauffman, M. D., R. D. MacArthur, M. D., Wm. J. Maynard, M. D., C. H. McGorray, M. D., W. T. Montgomery, M. D., W. H. Morgan, M. D., Daniel T. Nelson, M.D., John A. Robison, M. D., T. J. Shaw, M. D., W. A. Synon, M. D., H. M. Thomas M.D., A. E. Rhodes, M.D., E. J. Whitney, M. D. Visiting. Physicians.—F. H. Booth, M. D., C. E. Greenfield, M. D., Wm. B. Marcussan, M.D., C. H. McGorray, M. D., J. J. Tuthill, M. D. House Physician.—T. J. Dewey. Apothecary.—J. W. Ross. Superintendent.—Philip Adolphus, M. D.

Board of Directors.—President, Richard T. Crane, Esq.; Vice-President, John H. Williams; Secretary, Ephraim Ingals, M. D.; Treasurer, Joseph P. Ross, M. D.; Philip Adolphus, M. D., Jacob Beidler, Esq., Albert M. Billings, Esq., David Bradley, Esq., Norman Bridge, M. D., Elias Colbert, Esq., Chas. E. Culver, Esq., J. W. Farlin, Esq., L. R. Hall, Esq , E. L. Holmes, M. D., Jas. M. Horton, Esq., John W. Hutchins, M. D., John A. King, Esq., Geo. Mason, Esq., Henry R. Symonds, Esq., Hugh Templeton, Esq.

BENNETT FREE DISPENSARY.

At Bennett Medical College.

Clinics are held daily from 1 to 3 P. M., during the entire year. During the past year 3,500 cases were registered.

Scheme of Hospital and Dispensary Clinics.—Mondays, 1 to 3 P. M., H. G. Tucker, Venereal and Skin Diseases; Tuesdays, 1 to 3 P. M., S. D. Batcheldor, General Diseases; Wednesdays, 1 to 3 P. M., Clark, Diseases of Women, with operations; Thursdays, 1 to 3 P. M., McFatrick, Diseases of the Eye and Ear, with operations; Fridays, 1 to 3 P. M., Whitford, General Diseases; Saturday afternoons, Jay, General Surgical Diseases, with operatons.

LINCOLN STREET DISPENSARY.

335 LINCOLN STREET.

Dispensary Staff.—Mondays and Thursdays, Ellen H. Heise; Tuesdays and Fridays, Elizabeth H. Trout; Wednesdays and Saturdays, Marie J. Mergler.

NORTH STAR DISPENSARY.

At Emergency Hospital, 192 Superior street, between La Salle avenue and Wells street. Open for treatment of the sick of the deserving poor from 8 to 9 A. M., daily, except Sundays.

Medical Staff.—Surgeon, George H. Isham, attends Mondays, Wednesdays and Fridays Physician, J. M. Hall, attends Tuesdays, Thursdays and Saturdays; Oculist and Aurist, C. H. Beard, attends Mondays, Wednesdays and Fridays.

ARMOUR MISSION DISPENSARY,

SOUTHEAST CORNER OF THIRTY-THIRD AND BUTTERFIELD STREETS.

H. Martyn Scudder, Physician in Charge.

CHICAGO SPECTACLE CLINIC.

gives from 9 to 10 A. M., free of charge, a professional examination of the eye to those persons who cannot afford to pay for examination and glasses. Glasses given free of charge only on written order of the Chicago Charity Organization.

Rooms 208 and 209 Bay State Building, 70 State Street, Chicago. Dr. Fannie Dickinson, Surgeon in charge.

THE CHICAGO OPHTHALMIC COLLEGE DISPENSARY,

607 W. VAN BUREN ST., BET. MARSHFIELD AV. AND PAULINA ST.

Diseases of the eye, ear, nose, and throat treated daily, Sundays excepted, from 4 to 6 P. M. Eyes carefully tested for glasses.

· POLAND · SPRING · WATER ·

FROM

POLAND MAINE,

Is not only the PUREST and SWEETEST of drinking water, but is the SUREST CURE for all kinds of KIDNEY COMPLAINTS known. It is almost a positive cure, if taken freely, for the following diseases:

Bright's Disease of the Kidneys, Diabetes, Inflammation of the Bladder, Gravel, Urinary Calculi, or Stone in the Bladder, Stricture, Enlarged Prostate, and

CERTAIN FORMS OF ECZEMA.

President Garfield would drink no other Water.

[*Dispatch for Water from Hon. James G. Blaine.*]

WASHINGTON, D. C., Sept. 6, 1881.

H. S. OSGOOD, American Express, Portland, Me:

Send two more cases Poland to the President at Long Branch. Tell Mr. Ricker the President will drink no other water.

JAMES G. BLAINE.

☞ Send for 72 page Circular of the Poland Spring Water, giving a general history of the discovery of the Spring, and an analysis of the water.

H. RICKER & SONS, South Poland, Me., Proprietors.

SPRAGUE, WARNER & CO., Chicago, Ill., Western Agents.

CHICAGO PHYSICIANS.

Abbott, Geo. B. (R) Chi. Med., 1878; office, 163 State, 12 to 4; tel. 2480; res. 119 Madison, 10 to 12; tel. 1339.

Abel, John F. (R) Chi. Med. 1879; 3800 Dearborn, till 9, 1 to 3, and nights.

Adams, Carlos J. (R) Univ. Mich., 1868; office, 253 W. Madison, 2 to 4; res 856 W. Monroe; tel. 7227.

Adams, Charles (H) Hahnemann, 1872; 205 28th st., 8 to 9, 1 to 3; tel. 8151.

Adolphus, Philip (R) Univ. Md., 1858; office, 737 W. Madison, 2 to 3, 7 to 8; res. 636½ Washington boul., till 9; tel. 7056. Clinical Adjunct to Chair of Gynæcology, Rush Med. Col.; Gynæcologist Presbyterian Hospital; Med. Supt. Central Free Dispensary.

Akely, E. Seaman (E) Eclec. Med. Inst., Cincinnati, 1870; 199 W. Randolph, 9 to 11, 2 to 5; Sun. 8 to 10.

Akins, William T. (R) Med. Col., Fort Wayne, 1878; office, 362 Wabash ave., 9 to 12, 2 to 5, 7 to 8; tel. 22; res. 2353 Michigan ave.

Alderson, John J. (R) Chi. Med., 1885; office and res., 289 W. 12th, till 9, 1 to 3, nights; tel. 4378.

Alex, A. C. (R) Rush, 1883; office and res. 2910 Archer ave., 7 to 9, 3 to 8; tel. 8630.

Alexander, Harriet C. B., (R) Univ. of Mich , 1883; office, 21 Central Music Hall, 10 to 12; tel. 2642; res. Glencoe.

Allen, J. Adams (R) Castleton, 1846; office, 125 State, 2 to 5; res. 2001 Michigan ave. Pres. and Prof. of Principles and Practice of Medicine, Rush Med. Col.; Consulting Phys. Presbyterian and St. Joseph's Hospitals.

Allport, Walter W. (R) Rush; office, Argyle building, 10 to 4; tel. 2570; res. 69 Maple. Dental Surgeon St. Luke's Hospital.

Amet, Charles, (R) Office, 358 Ogden ave.; res. 958 Millard ave.

Anderson, P. E. Torgney (R) Rush, 1887; office, 128 Oak st., 9 to 10, 3 to 4, 7 to 8; Sun. 12 to 1; res. Elm st., near Market.

Anderson, Sarah A. (E) Bennett, 1883; 692 W. Adams.

Andrews, E. Wyllys, Chi. Med., 1881; office, 65 Randolph, 11 to 12:30; res. 6 E. 16th, 6 to 7; tel. 8242. Prof. Clinical Surgery, Chi. Med.; Surgeon, Mercy Hospital and South Side Free Dispensary.

Andrews, Edmund (R) Univ. Mich., 1852; office, 65 Randolph, 9:30 to 10:30; res. 6 16th, 12 to 1, 6 to 7; tel. 8242. Treas. and Prof. of Clinical Surgery, Chi. Med. Col.; Lecturer Ills. Col. of Pharmacy; Consulting Phys. Mercy Hospital, Chi.; Hospital for Women and Children, and Michael Reese Hospital; Surg. Dept. South Side Free Dispensary.

Andrews, Frank T. (R) Chi. Med., 1884; office, 65 Randolph, 4 to 6; tel. 5556; res. 6 16th; tel. 8242. Prof. Histology Chi. Med. Col.; Gynæologist Mercy Hospital.

Andrews, J. D. (R) Chi. Med., 1878; Surgeon's Office C. & N. W. Ry., 56 Kinzie, 8:30 to 9:30, 2:30 to 3:30; tel. 3106.

Andrews, Wells, (R) Rush, 1876; office and res., 979 W. Lake, till 9:30, 12 to 1:30, 6 to 7:30; tel. 7072.

Angear, J. J M. (R) Rush, 1860; office, 482 Lake, 1 to 3; tel. 7009; res. 143 S. Western ave., till 9, 11 to 12, 6 to 7; tel. 7059.

Armstrong, J. B. (R) New Orleans School of Med., 1869; office and res., 437 W. Madison, 9 to 11, 3 to 6; tel. 4090.

Arndt, Peter S. (R) Berkshire Med., 1842; office, Central Music Hall, 10 to 2; tel. 2642; res. 214 Cass, 6. p. m. to 8 a. m.

Arnulphy, Bernard S. (H) Faculty of Med., Paris, 1876; office, Central Music Hall, 12 to 2; tel. 2642; res. 344 La-Salle ave., 9 to 10, 6 to 7; tel. 3027. Prof. Physical Diagnosis Hah. Med. Col.

Atwood, Eugene S. (R) Rush, 1877; office, 113 E. Adams, 12 to 3; res. 2823 Calumet av., till 8, after 6.

Avery, S. J. (R) Rush, 1864; office, 25 Ashland block, 2 to 5; tel. 2453; res. 713 Washington boul., 8 to 9 a. m., 12 to 1:30; tel. 7128.

Axford, William L. (R) Univ. Mich., 1881; office, 70 Monroe, 2 to 6; tel. 5480; res. 52 31st, 11 to 12, 7 to 8; tel. 8354; Surgeon St. Joseph's Orphan Asylum.

B

Babcock, Elmer E. (R) P. & S., 1885; office, 1801 State, 8 to 9, 1 to 3, 7 to 9; tel. 8202; res. 1816 Indiana ave., 9 to 10, 3 to 4; tel. 8202 (Switch). Lecturer Surgical Anatomy at Col. P. & S.

Babcock, Robert H. (R) Chi. Med., 1878; P. & S., N. Y., 1879; office, 70 Monroe, 9:30 to 12:30, 3 to 5; tel. 80; res. 269 E. Erie. Prof. Diseases of Chest, Chi. Policlinic; Physician, Throat and Chest Dept. S. Side Free Dispensary.

Bachelle, G. V. (R) Rush, 1867; office, 322 Blue Island ave.; res. 18 Gilpin pl.

Bacon, C. S. (R) Chi. Med., 1884; office and res., Halsted and Webster ave., till 8, 12 to 4, 6:30 to 7:30; tel. 3120.

Bacon, Sara E. (H) Hahnemann, 1886; office and res. 171 E. 22d, 8 to 10, 4 to 5:30, 7 to 8; tel. 8165.

Bailey, E. S. (H) Hahnemann, 1878; 3034 Michigan ave., 8 to 10, 2 to 4, 7 to 8; tel. 8108. Clinical Diseases of Women, Hahnemann Hospital.

Bain, R. C. (H) Hahnemann, 1887; office and res., 3034 Michigan ave., 9 to 11:30, 7 to 8.

Baines, Oscar O. (E) Bennett, 1885; office and res., 657 Sedgwick, 7 to 9, 12 to 2, 7 to 8; tel. 3109. Demonstrator of Anatomy, Bennett Col.

Baldwin, A. E. (R) Rush, 1878; Chi. Dental Col., 1884; 828 W. Adams, 8:30 to 4:30, Wed. 1:30 to 5. Prof. Dental Pathology and Therapeutics in Dental Dept. N. W. Univ., Central Free Dispensary.

Ballard, E. A. (H) Hahnemann, 1863; office and res., 97 37th, 8 to 9, 1 to 3, 6 to 7.

Ballard, W. Harrison (R) P. & S. 1887; 119 Madison, 8 to 11, 3 to 7; tel. 1339.

Banga, Henry (R) Univ. of Basel, Switzerland, 1875; office, 70 State, 1 to 3; tel. 1339; office, 313 Sedgwick, 7 to 8 p. m.; tel. 3247; res. 456 LaSalle ave.; tel. 3237.

Banks, Jas. N. (R) Nat'l Med. Col. 1842; office, room 13, S. E. Cor. Clark and Washington, 10 to 11, 3 to 4; res. 2415 Michigan ave.

Barlow, Louis N. (R) Chi. Med., 1885; office, 2011 Clark, 9 to 10, 3 to 5, 7 to 8:30; tel. 8081; res. 2335 Wabash ave., till 8:30, 12 to 1:30; tel. 8461. Visiting Physician, S. Side Dispensary.

Barnum, Harry L. (H) Chi. Hom., 1883; Chi. Col. D. S., 1885; 628 W. Lake, 8 to 5; tel. 7016.

Barry, William T. (R) Bellevue Hosp., N. Y., 1884; 3229 Vernon av.

Bartholomew, James K. (R) Univ. Mich., 1887; office, 177 West Division, till 10, 1 to 2, 6 to 8; tel. 4573; res. cor. Milwaukee and Division.

Bartlett, John (R) Louisville, 1858; office and res., 480½ N. Clark, 11 to 12, 4 to 5; tel. 3063.

Bartlett, Rufus H. (R) Rush, 1879; office, 89 E. Madison, 1 to 4; tel. 7290; res. 116 S. Loomis, till 9, 5 to 7.

Bassett, Charles F. (H) Chi. Hom. Med., 1879; office, 3872 Cottage Grove ave., 11 to 12, 5 to 6; res. 20 Aldine sq.

Bassett, E. A. (E) Office, 109 Adams, 9 to 2, 5 to 8; res. 3546 Indiana ave.

Bassett, Geo. R. (E) Bennett, 1877; office, 1324 Ogden ave. and 761 Armitage ave.; res. 923 S. Homan ave.

Bassett, Susan A. (E) Bennett, 1884; office, 1324 Ogden ave., 9 to 12; res. 923 S. Homan ave.

Bates, Homer O. (R) Toronto, 1869; Col. P. & S., Keokuk, 1878; office, 527 S. Halsted, 9 to 10, 3 to 4; tel. 9139; office, 378 Blue Island, 1:30 to 2:30; tel. 9064; res. 802 S. Halsted, till 9, 11 to 1; tel. 9084. Prof. Diseases of Children, Chi. Policlinic.

Bates, J. Harvey (R) Long Island Col., 1874; office, 80 Madison, 10:30 to 12:30, 2:30 to 4:30; tel. 1453; office, 349 S. Clark, 9 to 10, 5 to 6, 8 to 9; res. 12 E. 16th.

Bates, Mary E. (R) Woman's Med. Col., 1881; office, 26 Central Music Hall, 2 to 3, except Wed.; tel. 5558; res. 527 W. Van Buren, 6 to 7, till 9 a. m.; tel. 7058. Prof. Anatomy, Woman's Med. Col.; Phys. Disp. and Hosp. for Women and Children.

Bausman, A. B. (R) Rush, 1882; 115 W. Madison; 9 to 10, 1 to 3, 6:30 to 8, evenings; tel. 4753.

Baxter, Andrew J. (R) Ohio Med. Col., 1860; office, 65 Randolph, 10:30 to 12, 3 to 5; tel. 5558; res. 334 W. Monroe, 1 to 2, 7 to 8; tel. 4521. Surg. Alexian Hosp., Cook Co. Hosp., Consulting Surg. Michael Rees Hosp.

Beach, George L. (H) Office, 103 Washington; res. 583 W. Harrison.

Beaman, Charles P. (H) N. Y. Hom., 1882; office, 163 State, 10 to 12, 1 to 4; tel. 2505; res. 54 Rush.

Beard, Charles H. (R) Univ. of Louisville, 1877; 70 Monroe, 9 to 1, 5 to 6. Asst. Surgeon, Ill. Charitable E. & E. Infirmary; Oculist and Aurist, N. Star Disp.

Bedell, Leila G. (H) Boston Univ., 1878; office and res., 181 Dearborn ave., 9 to 1, 6 to 7; tel. 3102.

Bedford, J. R. (R) Rush, 1881; office, 168 S. Halsted, 9 to 12, 3 to 6; tel. 4068; res. 424 W. Madison, till 8, 1 to 2, after 9; tel. 4090.

Bedford, L. (H) Philadelphia Homeo. Col., 1865; office and res., 3711 Ellis ave., 8 to 10, 4 to 6.

Beebe, Albert G. (H) Hahnemann, 1869; Bellevue, 1870; office and res., 552 W. Monroe, 8 to 9, 1 to 2, 6:30 to 7:30; tel. 7211. Prof. Operative Surgery, Chi. Homeo. Med. Col.; Surg. Cent. Homeo. Hosp. and Disp.

Beebe, Curtis M. (H) Chic. Homeo Med. Col., 1883; office and res. 23 S. Ashland ave., 8 to 9, 1 to 2, 6:30 to 7:30; tel. 7015. Prof. of Anatomy, Chic. Homeo. Med. Col.

Beebe, E. O. (H) Hahnemann, 1879; office and res., 139 29th.

Beery, Charles C. (R) Jefferson, 1879; office, 1801 State, 8 to 10, 1 to 3:30, 7 to 9; tel. 8202; res. 181 25th, till 8, 12 to 1, 5 to 7; tel. 8458.

Behrendt, Alex. (R) Berlin, 1883; office and res., 2505 Archer ave., 7 to 9, 5 to 7; tel. 8333.

Behrens, B. M. (R) Christiania Univ., Norway, 1875; office, 126 State, 2:30 to 4; office, 116 N. Center ave., 1 to 2, 6 to 7; res. 88 Park, 9 to 11.

Bein, William, (R) P. & S., 1887; office and res., 1475 Milwaukee ave.

Belfield, William T. (R) Rush, 1878; office, 612 Opera House bldg., 11 to 1; tel. 1298; res. 445 N. Clark, till 9:30, after 7; tel. 3148. Prof. of Physiology and Microscopy, Chic. Col. of Dental Surgery; Surgeon, Cook Co. Hospital; Attending Surgeon, Central Free Dispensary.

Bell, George, (R) Toronto Med. Faculty, 1888; res. 1410 35th, till 9, 12 to 2, after 7; office, 4222 Ashland, 3 to 5.

Bell, Henry (R) Univ. N. Y., 1881; 280 W. Indiana.

Bell, J. B. (R) Jefferson, 1861; office, 231 W. Madison, 9 to 10, 1:30 to 3:30, 7 to 9; tel. 4521; res. 21 Center ave., 12 to 1, 5 to 6.

Bell, J. J. (R) Rush, 1886; office, 858 Clybourn; res. 864 Clybourn ave.

Bemis, Joseph G. (E) P. & S., N. Y., 1865; Bennett, 1883; office, 103 State, 12 to 1; tel. 1339; res. 300 Maple ave., Oak Park. - Prof. Obstetrics, Bennett Med. Col.

Benson, John A. (R) P. & S., N. Y., 1880; office and res., 308 Chicago ave., till 7:45, 12 to 1:30, 7 to 8; tel. 3125. Prof. Physiology and Histology, Col. P. & S., Chi.; Prof. Physiology, Histology and Microscopy, Chic School Dermatology; Surg., St. Vincent's Foundling Asylum and Maternity Hosp.

Bergeson, John, (R) Univ. of Iowa, 1888; office and res., 1228 Milwaukee ave., till 9, 12 to 2:30, after 7; tel. 4203.

Bernard, Chas. C. (H) Chi. Hom. Med., 1882; office and res., 954 N. Halsted, till 10, 1 to 2, after 7; tel. 3374.

Berry, Charles H. (R) Univ. Mich., 1868; office, 103 State, 12 to 5; res. 24 Wisconsin; tel. 3201.

Berry, James G. (R) Rush, 1875; office, 3659 S. Halsted, till 9, 12 to 2, after 7; tel. 9589; res. 1515 E. 43d st., 3 to 4; tel. 9802.

Bersuch, Friedrich (H) Hahnemann, 1887; 82 Fullerton ave., 9 to 11, 2 to 4.

Bert, Ed (R) Univ. Goettingen, Germany, 1864; office, 52 31st, 2 to 4; tel. 8354; res. 3242 Vernon ave., 7 to 8; tel. 8004.

Besharian, John H. (R) Rush, 1882; office and res., 3160 State, 10 to 12, 3 to 4, after 7; tel. 8330. Physician, South Side Disp.

Bettman, Boerne (R) Miami Med. Col., 1877; office, Central Music Hall, 9 to 1, 4:30 to 5:30; tel. 2642; res. 3228 S. Park ave., till 9, 5 to 7; tel. 1071. Asst. Surg., Eye Dept. Ill. Charitable E. & E. Infirmary.

Bidwell, Theodore S. (R) Cleveland, 1863; office, 2906 Archer ave., 10 to 12; tel. 8195; res. 362 W. Jackson, before 9, after 5; tel. 4516.

Bigelow, Arthur W. (R) Col. P. & S., Ontario, 1885, Royal Col. of Phys., Eng., 1886; office and res., 3701 Cottage Grove, 8 to 9, 12:30 to 2:30, 6 to 8, night; tel. 9874.

Bigelow, Charles (R) 201 Clark, 9 a. m. to 8 p. m.

Bigelow, Jno. F. (R) Rush, 1882; office, 199 Clark; res., 723 W. Jackson.

Billings, Frank (R) Chi. Med. Col., 1881; office, 235 State, 11 to 2; tel. 2349; res. 65 23d, till 9, after 6; tel. 8616. Prof. Phys. Diagnosis and Clinical Med., Chic. Med. Col.; Visiting Surg., South Side Disp.

Bird, J. H. (R) P. & S., 1884; 240 Wabash; tel. 2810.

Birkhoff, D. (R) Rush, 1881; office, 547 Blue Island ave.; res. 647 W. Harrison.

Bishop, Rufus W. (R) Univ. of Berlin, 1882; office, 125 State, 10 to 12; tel. 5442; res. 2139 Wabash ave.; tel. 8133. Prof. of Phys. and Dermatology, Chic. Med. Col.

Bishop, Seth S. (R) Chi. Med., 1876; office and res., 719 W. Adams, till 10, 5 to 7; tel. 7218. Surg. to Ill. Charitable E. & E. Infirmary, to Ill. Masonic Orphans' Home, to West Side Free Dispensary.

Blake, S. C. (R) Harvard, 1853; office, 125 State, 9 to 12; tel. 2442; res. 576 Fullerton ave., 4 to 6; tel. 3967. Consulting Staff, Wom. and Chil. Hosp.

Blakeslee, L. K. (H) Hahnemann, 1883; 91 Warren ave.; 941 W. Harrison.

Blanchard, J. (R) Cincinnati Med. Col., 1846; office 34 Monroe, 12 to 3; tel. 5480; res. 177 N. State, 9 to 11, 7 to 9; tel. 3125.

Blanchard, Wallace (R) Chi. Med., 1869; office, 34 Monroe st., 9 to 11; tel. 5480; 239 State, 6 to 7, tel. 5349; res. 177 N. State, 4 to 5, night; tel. 3125.

Bliem, M. J. (H) Chi. Hom. 1884; Central Music Hall, 2 to 4; tel. 5241. Adj. Prof. Physiology and Demonstrator of Histology and Microscopy, Chi. Hom. Med. Col.

Blinn, Odelia (R) Wom. Med., 1868; office, 184 Dearborn, 12 to 1; tel. 112; res. 202 Warren ave., 8 to 10, 4 to 6; tel. 7049.

Blouke, M. B. (H) Chi. Hom., 1884; office and res., 193 Campbell ave., 8 to 9, 2 to 4, 7 to 8; tel. 7263.

Blunt, Arthur L. (E) Bennett, 1888; office, 136 Clark, 1 to 4; tel. 1339; res. 540 37th, 10 to 12, 7 to 10; tel. 9869.

Bluthardt T. J. (R) Chi. Med., 1861; office, Chi. Opera H. bldg., 2 to 4; res. 453 LaSalle, 8 to 9, 7 to 8.

Boas, Edmund A. (R) Med. Col., City of N. Y., 1881; Rush, 1884; office, 120½ Clybourne ave., 8 to 10, 7 to 9; tel. 3558; res. 46 Beethoven pl.

Bockius, F. B. Eisen (R) Chi. Med., 1872; offices, 234 Milwaukee ave., 10 to 11, 8 to 9; tel. 4011; 318 E. Division, 2 to 3; tel. 3284; res. 199 W. Indiana.

Boeber, F. W. (H) Office 625 Milwaukee ave., 10 to 4; res. Fullersberg, Ill.

Bogue, Roswell G.(R) P. & S., N.Y., 1857; office and res.,5 Washington place, till 9, 12 to 2, 6 to 7; tel. 3601. Consulting Surgeon, Presbyterian and St. Joseph Hosps., Chi. Hosp. for Women and Children and Central Free Disp.; Prof. Emeritus of Surg., Women's Med. Col.

Boice, Geo. W. (R) P. &. S., 1886; office and res., 1160 S. Western ave.

Bonynge, F. G. (R) Royal Col. Surg., Ireland, 1881; King's and Queen's Col. of Physicians, Ireland, 1880; office 70 E. Monroe, 9:30 to 1:30; tel. 5480; res. 307 N. Clark; tel. 3587.

Booth, Frank Hulburt (R) Rush, 1885; office and res., 402 S. Oakley ave., 11 to 1, 6 to 7; tel. 7006. Visiting Phys., Cent. Free Disp.

Borchsenius-Skov, H. S. (R) Women's Med., Penn., 1882; office and res., 73 N. Center ave., 10 to 12, 7 to 8; tel. 4235.

Borland, Leonard C. (R) Rush, 1887; office, 378 W. Van Buren, till 9, 1 to 3, and 6 to 9; tel. 4511; res. 365 W. Jackson. Supervising Phys., St. Luke's Hosp.

Borland, M. W. (E) Bennett, 1872; office, 378 W. Van Buren, till 9, 1 to 3, 6 to 7:30; tel. 4511; res. 365 W. Jackson.

Borter, F. X. (R) Univ. of Innsbruck, 1878; office, N. W. cor. Polk & Clark, 6 to 9, 1 to 6; tel. 1690; res. 4822 Ashland ave., 9:30 to 12; tel. 9543.

Bosworth, Alfred W. (R) Rush, 1868; 2979 Wabash ave., before 9, 1 to 3, evening; tel. 8145.

Bottomley, S. H. (R) Chi. Med., 1863; office and res., 342 Center st., 8 to 10, 6 to 8; tel. 3109.

Bouffleur, Albert I. (R.) Rush, 1887; Cook Co. Hosp.; tel. 7133; Interne Cook Co. Hosp.

Boughton, D. F. (R) Univ. of Mich., 1870; 321 W. Madison; 10 to 12, 3 to 6, 7 to 10; tel. 4521.

Boulter, H. H. (H) Hahnemann, 1880; N. W. Col. Den. Surgeons, 1886; office, 70 State; res. Ravenswood.

Bowen, Mary H. (R) Wom. Med., 1876; 30 Laflin, 10 to 12, 4 to 6.

Bowerman, Martha A. (H) Hahnemann, 1882; office and res., 1316 Oakwood boul., till 9, 1 to 4, and 6 to 7:30; tel. 9850.

Bowers, L. S. (R) Chi. Med., 1885; office, 721 Elston ave., 3 to 4; tel. 4804; res. 239 W. North ave., 7 to 9, 12 to 2, 5 to 7, nights; tel. 4774.

Bowlby, G. B. (R) Rush, 1888; office, 240 Wabash ave.; res. 1906 Milwaukee ave.

Boyd, Henry W. (R) Chi. Med., 1866; office and res., 3025 Indiana ave.; till 10 after 4.

Boyd, Laura E. (R) Indiana Med. Col., 1887; office and res., 360 Center.

Boyd, R. D. (R) Rush, 1878: office and res., 144 Oakwood boul.; 8 to 10, 1 to 3, and 7 to 8.

Boyer, Valentine A. (R) Univ. of Penn., 1836; office and res., 490 Fullerton ave.

Boynton, J. R. (H) Hahnemann, Phil., 1880; office, 70 State, 2 to 5; tel. 3530; res. 285 La Salle ave.

Bradbury, Robert (R) 854 Gross ave.

Bradley, Geo. F. (R) Rush, 1880; 287 W. 12th.

Bradley, James (R) Chi. Med., 1868; office and res., 70 Park ave.

Bradway, A. C. (R) P. & S., 1887; office and res., 1150 W. Harrison, till 8, after 4; tel. 7126.

Brand, Mathias (R) Chi. Med., 1886; office, 709 W. 21st, 2 to 4; tel. 9151; res. 145 Moore, till 10, 12 to 2, after 6.

Brandt, J. R. (R) Ohio Med. Col., 1875; office and res., 876 Clybourn ave., till 10, 1 to 3; tel. 12,024; Att. Phys. Cook Co. Hosp.

Braun, Louis, (R) Rush, 1876; office, 2603 Halsted, 11 to 12; res. 551 S. Halsted, till 9, 12 to 2, after 7; tel. 9030.

Brauns, Thilo (R) Royal Univ. of Berlin, Prussia; office and res., 825 N. Clark, 8 to 9, 2 to 4; tel. 3201.

Breckinridge, Stephen (R) St. Louis, 1879; 70 State, 2 to 3.

Breese, Ambrose (E) Bennett, 1887; office and res., 229½ Huron, cor. Clark; tel. 3439.

Breidenstein, F. L. (R) Strasburg, 1841; 789 Elk Grove ave.

Breiton, Nathan (R) P. & S., N. Y., 1888; office and res., 666 Carroll.

Brendecke, A. C. (R) P. & S., 1886; 56 W. Randolph; 7 to 9, 12 to 2, 6 to 8.

Brenneu, D. F. (R) Chi. Med., 1886; office, 348 Twenty-sixth; 10 to 12, 2 to 4, and 7 to 8; res. 3008 Butterfield.

Bridge, Norman, (R) Chi. Med., 1868; Rush; office and res., 550 W. Jackson, till 9, 12 to 1, 6 to 7; tel. 7038. Prof. of Pathology and Adj., Prof. Principles and Practice of Med., Rush Med. Col.; Attg. Phys., Presbyterian Hosp.

Bridge, W. C. (H) Chi. Hom., 1887; Cook Co. Hosp.

Brigham, B. A. (R) P. & S., 1886; office, N. W. cor. Randolph and State, 2:30 to 4:30; tel. 5558; res. 208 Milwaukee ave., till 10, 12 to 2, 7; tel. 4612. Diseases of Women Wed. and Sat., Gen. Medicine Tues. and Thur., W. Side Free Disp.; Lecturer on Anatomy, Col. P. & S.

Brill, John A. (R) Col. M. & S., Cincinnati, 1884; office and res., 428 Milwaukee ave., 8 to 10, 1 to 3, 6 to 8; tel. 4029.

Brinckerhoff, C. E. (R) P. & S., 1885; office and res., 182 N. Halsted, 8 to 10 and 2 to 4 and evening; tel. 4253.

Brinckerhoff, G. E. (R) Chi. Col. P. & S., 1885; office, 125 State; 9 to 12; tel. 2442; res. 121 S. Sangamon; Lecturer on Opth. and Otol., Col. P. & S.

Brindisi, Rocco (R) Univ. of Naples, 1884; office and res., 226 Milwaukee ave., 7 to 9, 12 to 2, 5 to 7; tel. 4011.

Brockway, Vira A (R) Woman's Med. Col., 1887; office and res., 547 S. Leavitt, 8 to 10, 4 to 6; tel. 7135.

Broe, R. H. (R) Chi. Med., 1878; office and res., 75½ 29th.

Broell, A. C. (R) Chi. Med., 1886; 83 Willow.

Brooker, J. B. (E) Eclectic Med. Soc., N. Y.; office and res., 278 Honore.

Brooks, Almon, (R) Univ. of Va., 1865; office, 240 Wabash av., 9 to 12, 2 to 5; tel. 8141; res. 2548 Indiana av.

Brooks, F. (E) Office, 245 Ogden ave.; res. 842 W. Jackson.

Brooks, J. W. (R) Jefferson, 1835; office and res., 2958 Indiana ave., 7 to 8, 1 to 2, 6 to 7; tel. 8282.

Brophy, T. W. (R) Rush, 1880; 96 State; res. 21 Bishop ct. Prof. Dental Pathology, Rush; Oral Surgery, Chi. Col. D. S.

Brougham, E. J. (R) Chic. Med., 1887; 663 N. Franklin.

Brower, Daniel R (R) Univ. of Georgetown, 1864: office, 70 State, 10:30 to 1; tel. 7085; res. 597 W. Jackson, till 8, 6:30 to 7:30. Lecturer on Practice of Med., Rush Med. Col.; Lecturer Ill. Col. Pharmacy; Med. Staff (mental and nervous diseases) St. Joseph's Hosp.; Consulting Phys. Woman's Hosp. and Wash. Home; Prof. of Diseases of the Nervous System and Clinical Medicine, Woman's Med. Col.

Brown, Harry (R) McGill, 1873; office, 70 State, 10 to 2.

Brown, Moreau Roberts (R) Louisville, 1876; office, 126 State, 10 to 12:30 and 5 to 6; Sunday, 11 to 12; tel. 624; res. 314 N. State, till 8:30 and after 7; tel. 3537.

Brown, I. W. (R) Rush, 1865; 873 W. Polk, 9 to 12, 1 to 5.

Bruce, Susan E. (H) Chic. Hom., 1880; 265 N. May.

Bruelly, James (R) Chi. Med., 1868; 70 Park ave.

Bryan, John C. (R) Rush, 1877; 1 Bryan pl.

Bryan, Mrs. Rose W. (R) Univ., Mich., 1886; office, 21 Central Music Hall, 2 to 4; tel. 2642.

Brydon, James M. (R) P. & S. Ind., 1875; office, 243 State, 11 to 12, 1 to 5, 7 to 8.

Bryson, W. G. (R) P. & S., N. Y., 1866; office, 70 State, 10 to 12 and 1 to 4.

Buchan, J. R. (R) Long Island, 1872; 627 W. Indiana, till 9, 1 to 2, after 7; tel. 7063.

Buchanan, C. H. (R) Rush, 1876; office, 3901 Cottage Grove ave., 9 to 10, 3 to 5, 7 to 8; Sun., 3 to 5.

Buchanan, Walter W. (R) P. & S., 1883; office, 134 Blue Island ave., 8 to 10, 2 to 4, 6 to 8; tel. 4101.

Buck, James P. (R) Jefferson, 1879; 418 LaSalle ave., till 9:30, 1 to 2, after 7.

Buecking, E. F. (E) Bennett, 1876; 561 W. 12th, till 9, 1 to 3 and 7 to 9. Prof. Surgical Anatomy and Orthopædic Surgeon, Bennett Col.

Buffum, Harry S. (H) Chic. Hom., 1886; office and res., 178 N. Franklin, 8 to 10, 1 to 3, 6 to 8; tel. 3274.

Buffum, Joseph H. (H) Hom. Med. Col., N. Y., 1873; office, 100 State, 9 to 12 and 4 to 5; res. 366 Ontario. Prof. of Diseases of Eye and Ear, Chic. Hom. Med. Col.; E. & E. Dept. Cen. Hom. Hosp. and Free Disp.

Burbank, W. M. (R) Castleton, 1844; office and res., 300 31st, 9 to 11, 2 to 5.

Burke, James (R) Rush, 1887; 908 W. 12th.

Burnett, Mary Weeks (H) Hahnemann, 1879; office, Central Music Hall, 1 to 4; res. 2500 Indiana ave.

Burnside, A. W. (E) Ec. Medical Inst., Cincinnati, 1853; 811 Washington blvd., 8 to 9, 12 to 1:30, 6:30 to 8.

Burr, Albert H. (R) Chic. Med., 1881; office, 279 State, 10 to 12, 2 to 5, 7:30 to 9; tel. 320; res. 1253 Wabash ave., till 9, 1 to 2, 6 to 7:30; tel. 8431.

Burry, James (R) Chic. Med., 1875; offices, 3718 S. Halsted; Rialto bldg.; res. 4012 Ellis ave.

Burt, W. H. (H) Cleveland Hosp. Col., 1858; office, 112 Dearborn, 9:30 to 10;30, 2 to 4; res. 652 W. Washington, till 9, 12:30 to 1:30, 6 to 7.

Burwash, Henry J. (R) McGill, Royal Coll. of Phys., London, 1879; office and res., 1218 Milwaukee ave., 8 to 9:30, 12 to 2, after 7; tel. 4203.

Bushee, G. B. (H) Hahnemann, 1887; 559 Chestnut, till 9, 12 to 2, 6 to 7; tel. 86.

Buttner, Adolph (R) P. & S., 1886; 52 Fremont, till 10, 12 to 2, 6 to 8.

Buzik, Julius (R) Vienna, 1885; 829 Milwaukee ave., 7 to 9, 1 to 3, 6 to 8; tel. 4573.

Byford, Henry T. (R) Chic. Med., 1873; office, 125 State, 1:30 to 3:30; tel. 5442; res. 3100 Forest ave., 8:30 to 9, 6:30 to 7; tel. 8299. Attg. Gynæcologist St. Luke's Free Hosp.; Med. Staff Michael Reese Hosp.; Lecturer on Obstetrics, Rush Med. Coll.

Byford, William H. (R) Med. Coll. of Ohio, 1844; office, 125 State, 9 to 12; tel. 5442; res. 1832 Indiana ave.; tel. 8283. Prof. of Gynæcology, Rush Med. Coll.; Lecturer, Ill. Coll. Pharmacy; Consulting Physician, Chic. Hosp. for Women and Children; Med. Staff, Woman's Hosp.; Consulting Phys., Cent. Free Disp. and Erring Woman's Refuge; Phys., Home for Incurables; Pres. and Prof. Gynæcology, Woman's Med. Coll.

Byrne, John H. (R) Rush, 1874; 231 W. Randolph, till 9, 12 to 1:30, 5:30 to 7:30; tel. 4270.

C

Cadieux, Joseph N. (R) office and res., 156 Blue Island ave.

Caldwell, Chas. (R) Harvard, 1867; office and res., 3237 Indiana ave., till 10, 12 to 1, after 7.

Caldwell, Charles Edwin (R) Rush, 1877; office, 3858 State, 9 to 10:30, 7:30 to 9; tel. 9824; res. 438 38th; tel. 9915.

Caldwell, Francis C. (R) P. & S., 1883; office, Central Music Hall, 2 to 6; tel. 2642; res. 631 Oakley ave.; tel. 7024.

Caldwell, W. C. (R) P. & S., 1885; office and res., 168 S. Halsted, 11 to 12, 7 to 8; tel. 4068.

Calkins, C. D. (R) Iowa State Univ., 1883; office, 225 Dearborn 10 to 12, 1 to 4; res 166 Dearborn ave., 8 to 9, 5 to 8.

Camp, A. Royce (R) Univ. Vermont, 1867; office, 147 4th ave., 10 to 5; tel. 1724; 444 Dearborn, 6 to 9:30; res. 3155 Wabash ave.

Camp, Chas. D. (R) Rush, 1878; office and res. 325 Washington blvd., 8 to 9, 2 to 3, 5 to 7; tel. 4656.

Canfield, Corresta T. (H) Cleveland Hom. Hosp. Col. and Woman's Med. Col., Cleveland, 1871; office and res. 244 Lincoln ave., till 9, 1 to 4; tel. 3269; Supt. of Clybourn Ave. Disp.

Carder, G. H. (H) Chi. Homeo. Med. Col., 1882; office and res. 368 Park ave., till 10, 12 to 2, after 4; tel. 7168.

Carey, T. L. (R) Chi. Med., 1868; office, 277 Clark, 2 to 4, 7 to 8; tel. 1077; res. 3613 Forest av., till 9, night; tel. 8590.

Carr, Abbie Underwood (H) Michigan Univ., 1874; Chi. Hom. Med., 1876; office and res. 318 Belden av., 9 to 12.

Carr, Wattson (R) Maryland Univ., Baltimore, 1848; office and res. 447 Washington blvd., 8 to 9, 11 to 12, 3 to 4, 7 to 9.

Carroll, J. G. (R) Univ. of Md., 1872; N. W. corner State and Washington, 9 to 11, 2 to 4, 7 to 9.

Carter, Chas. H. (R) Rush, 1880; office and res. 20 Ewing pl., till 8:30, 12 to 2, 7 to 8; tel. 4317; at cor. California and North aves., 4 to 5; tel. 4252.

Cary, Frank (R) Rush, 1882; office and res. 3027 Indiana ave., till 9, 12 to 1, 6 to 7; tel. 8108. Prof. Pathology Woman's Med. Col., Pathologist St. Luke's Hosp., Gynæcologist to St. Luke's Disp.

Cary, Harriet H. (R) Wom. Med., N. Y., 1883; office and res. 3027 Indiana ave., 11 to 1.

Case, Lafayette W (R) Rush, 1870; office and res. 384 N. Franklin, 9 to 10, 2 to 3; tel. 3301.

Casely, W. J. C. (R) Chi. Med., 1885; 3101 S. Halsted, 9 to 11, 2 to 4, 6 to 8; tel. 8061; res. 628 31st.

Casselberry, Wm. E. (R) Univ. of Penn., 1879; office, 70 E. Monroe, 3 to 5, 12 for messages; tel. 80; res. 3343 Calumet ave., 8 to 9, 6:30 to 7:30. Prof. of Materia Medica and Therapeutics and Laryngology and Rhinology, Chi. Med. Col.: Throat and Chest Dept., South-Side Free Disp.

Centaro, V (R) Univ. Naples, Italy, 1880; office, 329 S. Clark, 8 to 10, 2 to 4, 6 to 7; tel. 1401.

Cessna, Chas. E. (R) Rush, 1885; office and res. 1546 Milwaukee ave., till 9, 12 to 3, after 7; tel. 4729; branch office, 736 W. Division, 9 to 10.

Chaffee, John B. (R) McDowell Med. Col., St. Louis, 1861; office and res. 527 State, 9 to 11, 2 to 4, 7 to 9; tel. 1060.

Chamberlin, George M. (R) Rush, 1865; office and res. 3031 Indiana ave., 7 to 9; 4 to 6; tel. 8108.

Chambers, John E (R) Jefferson, 1877; 277 Clark, 9 to 1, 4 to 6, 7 to 9.

Chapin, Staley N. (R) Rush, 1885; office, 70 State, 10 to 1, 2 to 4; tel. 5558; res. 2237 Prairie ave. Attg. Phys. Central Free Disp.

Chew, John H. (R) Univ. Md., 1863; office, 211 Chicago Opera House, 3 to 5; res. 417 N. State, till 8:30, 1 to 2; tel. 3296.

Christie, Edmund (R) McGill, 1882; 681 W. 21st; 573 Blue Island ave.

Christoph, E. O. (R) Univ. Freiburg, Germany, 1887; office and res. 390 35th, 12 to 2, 5 to 7; tel. 8066.

Church, Nelson H. (R) Rush, 1869; office, 802 S. Halsted, 9 to 11, 2 to 4, 6 to 8; tel. 9084; res. 1161 Filmore.

Clark, Anson L. (E) Eclectic Med. Inst., Cincinnati, 1861; office, 511 State, 2 to 4, Wed. and Sat.; tel. 667; res. Elgin. Prof. Obstetrics and Diseases of Women and Clinical Gynæcology, Bennett Col.; Gynæcologist, Bennett Hosp.; Prof. Diseases of Women, Bennett Free Disp.; Mem. State Board of Health.

Clark, Charles M. (R) Univ. City of N. Y., 1857; office and res., 1085 Grenshaw, till 10, 4 to 6; tel. 7024.

Clark, J. A. (R) Chi. Med., 1885; office and res. 900 W. 21st, 8 to 9, 12 to 2, 6 to 7:30; tel. 9204.

Clark, J. S. (R) Geneva, 1843; 558 LaSalle ave., 9 to 10, 6 to 8; tel. 3027.

Clark, Jennie B. (R) Wom. Hosp. Med., 1886; office and res. 900 W. 21st, 12 to 2, 6 to 8; tel. 9204.

Clark, T. A. (R) Univ. of Louisville, 1852; office and res. 256 S. Halsted, 1 to 3, 7 to 9, nights; tel. 4067.

Clark, Wallace C. (R) Univ. N. Y. City, 1882; office, 194 Michigan ave., 9 to 11, 3 to 5.

Clarke, Albert L. (R) Bellevue, 1878; office and res., 388 35th, till 10, 1 to 2, 6:30 to 7:30; tel. 8066.

Clarke, Franklin F. (H) office, 186 Clark, 8 to 8; Sunday, 9 to 12.

Clarke, O. H. E. (R) McGill, 1870; office and res., 3857 Vincennes ave. and 202 39th, 9 to 10, 3 to 4; 7 to 9, Tues., Wed. and Fri.; Sun., 1 to 2.

Clarke, Ward Greene, (R) Rush, 1882; office, S. E. corner Clark and Huron. 10 to 12, 4 to 6, 7:30 to 8; Sun., 3 to 5; tel. 3563; res. LaSalle ave., cor. Locust, till 9 a. m. and night; tel. 3342; Prof. Surg. Am. Col. Dental Surg.

Clarke, William E. (R) Vermont Med. Col., 1845; 690 W. Monroe, 8 to 9, 1 to 3. Consulting Phys. Chi. Hosp. for Women and Children.

Clary, W. J. (E) Eclectic Med. Col., Cincinnati, 1852; office, 163 State 11 to 5; res. Maplewood, till 10, evening.

Clendenen, I. (E) Bennett, 1872; office, 78 State, 10:30 to 11:45; res. Maywood. Prof. Oral Surgery, Am. Col. Den. Surgery.

Clevenger, Shobal Vail, (R) Chi. Med. 1879; office, 29 Central Music Hall, 10 to 3; tel. 5558; res. 245 Lincoln ave., after 5 p. m.; tel. 4208. Consulting Phys. Nervous and Mental Diseases, Michael Reese and Alexian Bros.' Hospitals.

Cobb, Jos. P. (H) Hahnemann, 1883; office, 207 31st, 9 to 10, 3 to 4, 7 to 8; res. 3638 Indiana ave., till 9, 1 to 2; tel. 8050. Consulting Phys. Home for Friendless; Phys. in charge Children's Clinic, Hahnemann Hosp.; Adj. Prof. Physiology and Histology Hahnemann Col.

Coe, Milton F. (R) P. and S., 1888; office, 144 Oakwood boul., 10 to 12, 3 to 5; tel. 9966; res. 221 E. 42nd; tel. 9913.

Coey, Andrew J. (R) Chi. Med., 1880; State and 31st, 9 to 10, 3 to 5, 7 to 8; tel. 8166. Phys. Cook County Hospital.

Coffin, W. V. (R.) Miami Med. Col., Cincinnati, 1880; office and res., 3545 Vincennes ave., 9 to 12, 3 to 2.

Coker, William Wilson (R) London Royal Col , Eng., 1872; Col. P. & S., Chi., 1888; office and res., 2802 Archer ave., 8 to 11, 2 to 4, 6 to 10.

Colburn, J. Elliott (R) Albany, 1877; office, 126 State, 11 to 2; tel. 1339; res., 489 Monroe, 6 p. m. Prof. Ophthalmology and Otology, Chi. Policlinic; Phys. Cook Co. Hospital, Eye and Ear Infirmary, Central Free Dispensary.

Colby, B. D. (R) Rush, 1884; office and res., 285 Loomis, till 8, 12 to 2, 6 to 7; tel. 4652.

Cole, Samuel (R) Rush, 1865; 3305 Vernon ave.; tel. 8019.

Coleman, W. Franklin (R) Royal Col. P. & S., Canada, 1863; office, 163 State, 9:15 to 12:15, 4:30 to 5:30; Sunday, 1 to 1:30; tel. 5480; res. 3132 Rhodes ave., 1 to 1:30; Sunday, 2 to 2:30; tel. 8354.

Collier, A. W. (H) Hahnemann, 1888; office and res., 3249 LaSalle, 8 to 10, 6 to 8.

Collins, D. (R) Univ. N. Y., 1880; office and res., 449 26th, 8 to 10, 1 to 3, 6 to 8; tel. 8225.

Colton, Adolphus D. Office and res., 1601 State, 9 to 12, 2 to 5.

Colton, W. W. (R) Rush, 1885; 47 Kinzie, 9 to 12, 2 to 4, 7 to 8; tel. 3092.

Colwell, Benj. L. (H) Hahnemann, 1882; 3014 Calumet, till 11; tel. 8432.

Condit, Gertrude Scott (R) Woman's Med., Baltimore, 1884; office, 102 S. Halsted, 10 to 1; res., 877 Washington boul.; tel. 7195.

Conley, P. H. (R) Rush, 1886; office and res., 170 Ashland ave., 9 to 10, 1 to 2, 7 to 9; tel. 7157.

Connolly, Geo. B. (H) Chi. Hom. Med., 1887; 73 Fullerton ave.

Constant, F. M. (R) Rush, 1856; 279 N. Market; tel. 3342.

Cook, H. H. (R) Univ. Mich., 1872; office and res., 27 N. Clark, till 10, 1 to 3, 7 to 8, and nights.

Cook, J. F. (R) Office, 81 Clark, 9 to 12, 2 to 4, 7 to 8; res., 517 Carroll ave.

Cook, W. B. (E) Bennett, 1882; 3131 Archer ave.; tel. 9605.

Cooke, Alexander Hardy (R) Univ. of N. Y., 1846; Victoria Med. Col., Canada, 1866; Col. P. & S., Canada, 1866; office and res., 234 Dearborn ave., till 8, 12 to 2, 6 to 8; tel. 3002. Phys. Home for Incurables.

Cooke, J. Masson (R) Rush, 1868; 234 Dearborn ave.; tel. 3002.

Coop, B. D. (R) Rush, 1871; office, 126 Jackson, 9 to 11:30, 3 to 5, evenings; res., 246½ S. Sangamon.

Copeland, W. L. (R) McGill, Montreal, 1872; Royal Col. Surg. Eng., 1873; office, 163 State, 3 to 5; tel. 5480; res., 866 W. Monroe, 1 to 2, 7 to 8; tel. 7040. Prof. Anatomy Chi. Col. Dent. Surg.; At'g. Phys. W. Side Free Disp.

Cotton, A. C. (R) Rush, 1878; office, 240 Wabash ave., 11 to 1; tel. 5457; res., 193 S. Wood, till 10, after 6; tel. 7274. Lecturer on Therapeutics, Rush Med. Col.; Phys., Cook Co. Hospital.

Coulter, H. A. (E) Bennett, 1881; office, 103 State, 11 to 3; res., 815 Washtenaw ave.

Cowgill-Bates, Laura (R) Woman's Hosp. Med., Chi., 1885; office and res., 802 S. Halsted, 2 to 4; tel. 9084.

Craig, James D. (H.) Hygieo-Therapeutic Col. of N. Y., 1858; office, 16 and 17 Central Music Hall, 9 to 11; tel. 5384; res., Rogers Park, till 8, 4 to 6; tel. Rogers Park Tel. Sta.

Crawford, S. K. (R) Univ. Mich. 1861; office and res., 680 W. Monroe, till 9, 12 to 2, 6 to 7. Prof. Surg. Anatomy Col. P. & S., Chi.; Attg. Gynæcologist (Wed. and Sat.) W. Side Free Disp.

Creighton, Charles J. (R) Rush, 1879; office and res, 2840 State, till 9, 3 to 5, after 7:30; tel. 8185.

Cronin, P. H. (R) Missouri Med. and St. Louis Univ., 1878; office, 501 and 503 Opera House bldg., 12 to 2; tel. 1294; res. 468-470 N. Clark, 9 to 11, 6 to 7:30; tel. 3312.

Cross, E. D. (R) Col. P. & S., Baltimore, 1878; office, 103 State, 10 to 3; tel. 8108; res. 3151 Indiana ave., till 9:30, 4 to 6.

Cunningham, George P. (R) N. Y. Coll. of Pharmacy, 1868; Rush, 1877; 237 N. Clark, 9 to 11, 1 to 3, 6 to 9; tel. 3002.

Curran, Patrick (R) Rush, 1867; 237 W. North ave., 10 to 12, 2 to 4; res. 62 Homer.

Curtis, J. H. (R) Univ. City of N. Y., 1885; office, 802 S. Halsted, 10:30 to 12; tel. 9084; 378 Blue Island ave., 12:30 to 1:45; tel. 9064; res. 440 W. Harrison, 8 to 10, 2 to 4, 6 to 8; tel. 4094.

Curtis, Lester (R) Chic. Med., 1870; 35 University pl., 9 to 11.

Cutler, George A. (R) Univ. Med. Coll., N. Y. City, 1853; office, 193 S. Clark; tel. 1339.

Cyrier, C. E. (R) Rush, 1879; office, 173 Blue Island ave., 8 to 10, 2 to 4, 7 to 8; res. 107 Blue Island ave.

D

Dahlberg, Alfred (R) Rush, 1881; office, 2459 Wentworth ave., 10 to 12:30, 6 to 8; tel. 8187; res. 5727 Wentworth ave., 8 to 9:30; tel. 95, Englewood.

Dal, Jacob (H) Hahnemann, 1871: office and res., 142 Evergreen ave., till 9, 12 to 2, after 6; tel., 4312.

Dal, John W. (R) Chic. Med., 1878; office and res., 860 Milwaukee ave., till 9, 12:30 to 2, after 6;30; tel. 4808.

Damon, O. B. (R) Harvard, 1866; office and res., 274 Park ave., 8 to 9, 1 to 2, 6 to 7; tel. 7048.

Danforth, I. N. (R) Med. Dept. Dartmouth Coll., 1869; office, 69 Dearborn, 11 to 1; tel. 2453; res. 294 W. Monroe, till 9, after 6; tel. 4131. Prof. of Clinical Med., Chic. Med. Coll,; Prof. Gen. Med., Chic. Training School; Lecturer Ill. Coll. of Pharmacy; Att'g Phys. St. Luke's Free Hosp.; Consulting Phys. Ill. Charitable E. & E. Infirmary; Prof. of Renal Diseases, Woman's Med. Coll.

David, Cyrenius A. (R) Rush, 1869; Univ. of N. Y., 1882; office, 125 State, 9 to 12; tel. 5442; res. 332 Dayton; tel. 3374.

David, J. C. (H) Hahnemann, 1876; office, 103 State, 2 to 4; tel. 5505; res. 1421 Oakwood boul., till 9 a. m.; tel. 9850.

Davidson, Charles (R) Chi. Med., 1883; office and res. 1002 Madison, 11 to 12, 3 to 6; tel. 7057. Ear Dept., Ill. Eye and Ear Infirmary.

Davidson, W. M. W. (H) Hahnemann, 1878; office, 565 W. Madison, 11 to 4; tel. 7172; res. 658 Walnut, till 10, after 5; tel. 7007.

Davis, Charles Gilbert (R) Eclectic Med. Inst., Cincinnati, 1871; Univ. of Virginia, 1873; Missouri Med. Coll. 1874; office, 240 Wabash ave.; tel. 5457; res. 705 W. Adams.

Davis, J. J. (R) Rush, 1885; office and res., 169 W. Madison, till 10, 2 to 5, 7 to 9; Haymarket tel.

Davis, L. J. (R) Ohio Med. Coll., Cincinnati, 1881; office and res., 552 W. Madison, 9 to 11, 2 to 4, 6 to 8; tel. 7022.

Davis, N. S. Jr. (R) Chic. Med., 1883; office, 65 Randolph, 10 to 12, 2 to 4; res. 326 Superior. Prof. Principles and Practice Med., Chic. Med. Coll.; Ass't attending Phys., Mercy Hosp.

Davis, N. S., Sr., (R) Col. of P. & S., N. Y., 1837; office, 65 Randolph, 9 to 2; tel. 5558; res. 291 Huron. Prof. Principles and Practice of Medicine and of Chemical Medicine, and Dean Chi. Med. Col.; Senior Phys. Mercy Hosp.

Davis, T. A. (R) P. & S., 1885; office and res. 977 Van Buren, till 10, after 4; tel. 7126. Surg. West Side Free Disp.; Lecturer on Surgery Col. P. & S.

Davis, W. H. (E) Eclectic Med. Inst., Cincinnati, 1865; office, 126 State, 10 to 4; tel. 624; res. 3423 Indiana ave.

Deal, D. B. (R) Office, 70 State, 10 to 2; res. 46 S. Morgan.

Deans, E. (R) N. W. Univ. Chi., 1863; office and res. 1800 Wabash ave., 11 to 2; tel. 8319.

De Bausset, Arthur (R) Sucre, Bolivia, 1861; 236 State.

De Bey, William (R) Rush, 1871; office and res. 651 W. Monroe, 8 to 10, after 6.

Deegan, William (R) P. & S., N. Y., 1884; office and res. 528 W. Indiana, 8 to 10, 2 to 4, nights; tel. 7169.

Deimel, Henry L. (E) Bennett, 1888; office and res. 2719 Indiana ave., 10 to 4.

Delamater, N. B. (H) Hahnemann, 1873; office 125 State, 9 to 2; tel. 5444; res. 3912 Lake ave. Prof. of Mental and Nervous Diseases, Chi. Homeo. Med. Col.; Supt. Cent. Homeo. Hosp. and Free Dispensary.

DeMarr, George E. (R) Western Reserve, 1875; office, 262 S. Halsted, 8 to 11, 2 to 4; tel. 4067; res. 447 Ogden ave.; tel. 7177.

Deming, H. H. (R) Chi. Med., 1878; office, 144 Oakwood blvd., 10:30 to 12:30, 3 to 5; tel. 9850; res. 3602 Vincennes ave., 7 to 8, 1 to 2, 7 to 7:30; tel. 9831.

DePuy, A. H. (R) Rush, 1856; office, Grand Pacific; res. Pullman building.

DeSauchet, A. L. (H) Chi. Hom. 1885; 70 State, 11 to 5.

DeVerry, Stephen C. (R) Univ. of Penn., 1871; office, 802 S. Halsted, 4 to 5; tel. 9084; res. 2542 Indiana ave., 1 to 3, 6 to 8; tel. 8169.

DeVries, F. P. Office, 277 Sedgwick, 8 to 11, 4 to 8; res. 374 Division, before 8 a. m., after 8 p. m.

Dewey, Charles A (H) Hahnemann, 1881; 207 31st st., till 9, 1 to 3, 7 to 8; tel. 8108.

Dewey, F. J. (R) Rush, 1885; office, 126 State, 5 to 7; tel. 1339; res. 688 W. Madison, 11 to 1; tel. 7270.

DeWitt, John H (R) Jefferson, 1880; office, 271 Wabash ave., 9 to 12, 2 to 5; res. 106 S. Centre ave.

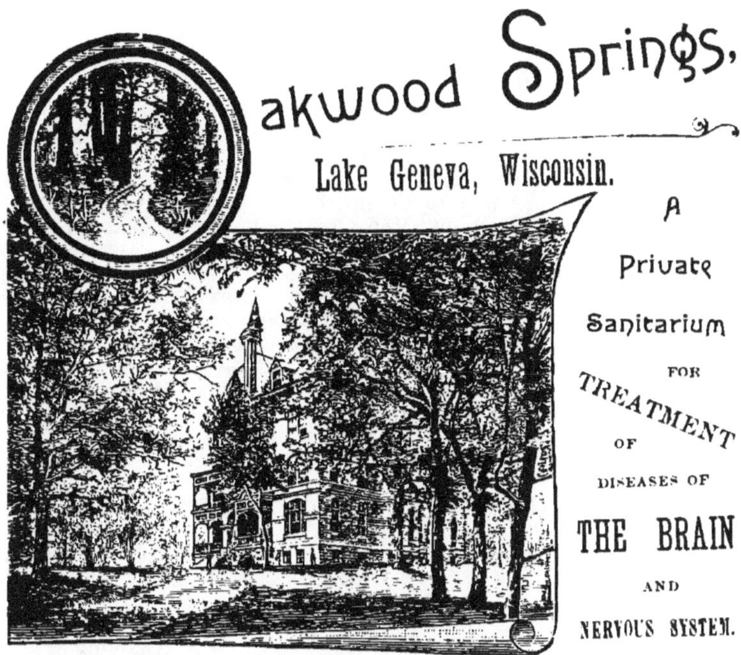

MEDICAL STAFF:

OSCAR A. KING, M.D., *Superintendent and President*, Professor of Diseases of the Brain and Nervous System, College of Physicians and Surgeons, Chicago; Physician-in Chief to Department for Nervous Dis ases, West Side Free Dispensary, Chicago; Professor of Psyciatry and Neurology and of Medical Electricity, Chicago Ophthalmic College.

LOUIS R. HEAD, M.D., *Resident Medical Superintendent.*

ATTENDING PHYSICIANS.

W. H. BYFORD, M. D., Chicago, Ill., Professor of Diseases of Women, Rush Medical College, Chicago.

HENRY PALMER, M.D., Janesville, Wis., Surgeon-General of Wisconsin, and Professor of Operative and Clinical Surgery, College of Physicians and Surgeons, Chicago.

WILLIAM E. QUINE, M.D., Chicago, Ill., Professor of Practice of Medicine, College of Physicians and Surgeons, Chicago.

WM. T. BELFIELD, M.D., Chicago, Ill., Pathologist of Cook County Hospital, Professor Genito-Urinary Organs, Chicago Polyclinic.

JOHN E. HARPER, M.D., Chicago, Ill., Professor of Ophthalmology and Clinical Diseases of the Eye and Ear, College of Physicians and Surgeons, Chicago.

CONSULTING PHYSICIANS.

A. REEVES JACKSON, M.D., HENRY M. LYMAN, M.D., D. R. BROWER, M.D., Chicago, Ill.

See pages 16, 137, and others.

DeWolf, Oscar C. (R) Univ. City N. Y., 1858; office, Health Dept., City Hall, 10 to 5; tel. 447; res. 435 Washington blvd. Prof. State Med. and Pub. Hygiene, Chi. Med. Col.

Dexter, Ransom (R) Univ. of Mich., 1862; office, 76 E. Madison, 8 a. m. to 9 p. m.; res. 2920 Calumet ave.

Dickerson, Louise (R) Wom. Med., 1881; office, 241 Wabash ave.; res. 122 21st.

Dickinson, Fannie (R) Woman's Med., 1883; 70 State, 9 to 12, afternoon hours by app. Pres. Woman's Col. of Optics and Ophthalmology; Head Phys. and Surg. Chi. Spectacle Clinic.

Dickson, Guy B. (H) Hahnemann, 1886; office and res. 1001 W. Harrison, 8 to 9, 1 to 3, 7 to 8; Sun., 9 to 10, 2 to 4.

Dietrich, Henry (R) Swiss Med. Council, Zurich, 1873; office, 95 5th ave., 1 to 2; res. 312 Sedgwick, 8 to 9, 7 to 8; tel. 3247.

Dilliard, E. (H) Hahnemann Med. Col., 1887; 129 22d, 8 to 10, 2 to 4, 7 to 8.

Dion, Delvina (R) Woman's Med., 1886; office, 240 Wabash ave., 2 to 5; tel. 5457; res. 36½ Vernon Park pl., till 10 and at 6; tel. 4559.

Dobbin, Annette S (R) Wom. Hosp. Med., 1882; office and res. 821 Warren ave. Asst. to Chair of Anatomy Wom. Med. Col.

Dodds, W. E. (R) Rush Med. Col., 1885; office, 89 Randolph, 12 to 8; res. cor. 57th st. and Jefferson ave., nights and till 11:30 a. m.; tel. 9933.

Dodge, C. C. (H) Chi. Hom. Med., 1887; office and res. 726 Washington blvd; tel. 7305.

Doe, A. (R) Univ. of Christiana, Norway, 1878; 284 W. Indiana st., 9 to 10, 3:30 to 4:30, 7 to 8: Sun., 12 to 1; tel. 4254.

Doering, E. J. (R) Chi. Med., 1874; office and res. 2406 Prairie ave., 1 to 3; tel. 8248.

Donoghue, Elizabeth B. (H) Hahnemann, 1881; office and res., 323 Chicago ave.

Doolittle, William H. (R) Rush, 1876; office and res., 165 18th, 10 to 12.

Dorland, E. H. (R) Miami, 1880; office, Opera House bldg., 2 to 4; res., 4329 Lake av., till 9, after 5; tel. 9982.

Dorsey, Nicholas James (R) Univ. of Maryland, 1847; office and res., 281 Erie, till 9, 12 to 2, 6 to 9; tel, 3439.

Downer, C. L. (R) Rush, 1884; office, 46 S. Clark, 10 to 1, 2 to 6; Sundays, 2 to 4; tel. 2201; res., 1020 W. Monroe, till 9; tel. 7057.

Downs, Charles M. (R) Yale, 1883; office, 681 Larabee; res. 499 La-Salle ave.

Dudley, E.C. (R) Long Island Hosp. Col., 1875; office, 70 Monroe, 11 to 12; tel. 80; res. 1619 Indiana ave.; tel. 8514. Prof of Gynaecology Chi. Med Col.

Med. Staff, Mercy Hosp.; Attg. Gynæcologist St. Luke's Hosp. and S. Side Free Disp.; Consulting Phys. Errring Woman's Refuge.

Duncan, D. (H) Hahnemann, 1887; office, 56 State, 11 to 3; Sun., 2 to 4; tel. 7214; res., 1084 W. Harrison, till 9:30, after 6.

Duncan, T C. (H) Hahnemann, 1866; office, 100 State, 11 to 3; res., 590 W. Adams, till 9, and 6 to 8.

Durselen, Charles N. (R) Univ. of Bonn., 1865; Univ. of Cincinnati, 1873; office and res., 2524 State, 9 to 11, 2 to 4, 7 to 7:30; Wed , 8 to 8:30 p. m.; Sun., 9 to 11, 2 to 4; tel. 8117.

Dyas, G. K. (R) Chi. Med., 1869; office, 107 Clark, 2 to 5; tel. 1206; res., 30th and Michigan ave.

Dyas, W. Godfrey (R) Royal Col. Surg., Ireland, 1845; office, 107 Clark, 9 to 1. Prof. Emeritus Theory and Practice of Med., Woman's Col.; Consulting Phys. Woman's and Children's Hosp.; Consulting Surg. St Joseph's Hosp.

E

Earle, Charles Warrington (R) Chi. Med., 1870; office and res., 535 Washington boul., 10 to 11, 6 to 7; tel. 7149. Prof. of Obstetrics, Col. P. & S., Chi.; Prof. of Diseases of Children, Chi. Training School; Lecturer Ill. Col. of Pharmacy; Phys. Washingtonian Home; Prof. of Diseases of Children and Clinical Med., Woman's Med. Col.

Earle, Frank B. (R) P. & S., 1885; office and res., 535 Washington boul., 10 to 11, 7 to 8; tel. 7149. Lecturer Prac. of Med., Col. P. & S.

Earll, Charles 328 W. Van Buren.

Earll, C. A. Cook Co. Hosp.

Eaton, David B. (R) Chi Med., 1882; office and res., 2725 State, 8 to 9, 1 to 2, nights; tel. 8520.

Edgar, W. H. (E) Bennett, 1880; 3901 State, 9 to 11, 3 to 5.

Edwards, E. W. (R) Washington, 1846; office, Opera House bldg., 11 to 1, 2 to 7.

Edwards Samuel (R) Office, 199 Clark; res., 1734 Wabash ave.

Egan, D. (R) Rush, 1881; office and res., 2899 Archer ave., 7 to 9, 10 to 12, 1 to 5, nights; tel. 8297.

Egbert, James K. (R) Rush Med. Col., 1886; office and res., 471 W. Madison, 1 to 3, nights; tel. 7022.

Eggleston, W G. (R) P. & S., N. Y., 1881; office, 65 Randolph; res., Clarendon Hotel.

Ehlers, P. F. T. 109 Adams.

Eldred, Wm. H. (R) Rush, 1882; office, cor. Kinzie and Market; tel. 3092.

Eldridge, C. S. (H.) Hom. Med., N. Y., 1868; office, 70 State, 9 to 11, 2 to 4; tel 2211; res., Tremont House.

Ellsberry, Isaac N. (R) Office and res., 3424 Dearborn, 9 to 11, 1 to 3, 7 to 8.

Ellingwood, Finley E. (R) Bennett, 1878; office, 3656 State, 11 to 12, 4 to 5, 7 to 8:30; tel. 9869; res., 3926 Indiana ave., till 9, 12 to 3; tel. 9941. Prof. of Hygiene and State Med. and Demonstrator of Analytical Chem., Bennett Med. Col.

Elliott, W. W. (R) Missouri Med. Col., 1875; office and res., 236 Sheffield ave., till 9, 12 to 2, after 5; tel. 3299.

Ely, C. F. (H) Hom. Med., N. Y., 1877; office, 103 State, 1 to 3; res., 277 Ontario.

Emery, R (R) Victoria, Univ., Canada, 1880; 125 S. Clark, 9 to 12, 1 to 5, 7 to 9.

Emmons, F. A. (R) Rush, 1863; office and res., 3422 Indiana ave., till 9, 2 to 4.

Emrick, G. M. (R) Chi. Med., 1873; 467 W. Chicago ave., 7 to 9, 1 to 3, 6 to 8; tel. 7250.

Engert, Rosa H. (R) Wurzburg, 1863; Wom. Med., 1873; office, Central Music Hall, 2 to 5; res. 486 N. Clark.

Epler, Ernest, (R) Chi. Med., 1883; 505 Canal, 10 to 11, 3 to 4, 7 to 8; tel 4494.

Ermine, Lucy E. (R) Wom. Hosp., Med., 1884; office and res., 747 Warren ave., 2 to 5; tel. 7007.

Etheridge, James H. (R) Rush, 1869; office, 65 Randolph, 10:30 to 1; tel. 5598; res. 1634 Michigan ave., till 8, 6 to 7; tel. 8272. Prof. Mat. Med. and Therap. and Med. Jurisprudence, Rush; Gynæcologist to Presbyterian Hosp.; Attending Phys. St. Luke's Hosp.; President Chi. Med. Society.

Evans, Charles Horace, (H) Hom. Med. of Penn., 1869; office, 570 W. North ave., 11 to 12; res. 730 Warren ave., 8 to 9, 6 to 7; tel. 7007.

Evans, Charles W. (R) Rush, 1886; 341 Fulton, 8 to 9, 12 to 1, after 7; tel. 4193; 126 Milwaukee ave., 10 to 12, 4 to 5; tel 4288.

Evatt, W. H. P. (R) Dublin, 1876; Royal Surg., Ireland, 1879; office, 547 Blue Island; res. 730 W. 21st.

Everett, Edward, (H) Hahnemann 1882; office, Central Music Hall, 12 to 3; res. 1130 N. Halsted.

Ewell, Marshall D. (R) Chi. Med. Col., 1884; office, 97 Clark, 9 to 12.

Ewing, Alice, (H) Hahnemann, 1887; 144 Oakwood boul., 9 to 12, 7 to 8; tel. 9966.

F

Falls, S. K. (R) McGill, 1875; 1028 W. Monroe, 8 to 10, 2 to 4, 7 to 9; Sun., 10 to 12; tel. 7057.

Farley, I. P. (R) Rush, 1885; 269 Bissell.

Farr, A. L. (R) Rush, 1880; 138 Lincoln ave., 8 to 10, 4 to 5, after 8; tel. 3634.

Feder, Henry C. (R) McGill, 1881; office, 100 S. Halsted, 3 to 4, 7 to 9; Sun., 2 to 4; res. 356 Washington boul., till 8, 1 to 2:30, nights; tel. 4827.

Fellbrich, John, (R) Office and res. 62 Lytle.

Fellows, H. B. (H) Western Col. Homeo. Med., 1861; 2969 Indiana ave., 8 to 10, 2 to 4; tel. 8113. Prof. Principles and Practice of Med., Hahnemann Med. Col.

Fenger, Christian, (R) Univ. of Copenhagen, 1867; office, 125 State, 2:30 to 4; tel. 5442; res. 269 LaSalle ave., till 9; tel. 3664. Prof. of Clinical Surg., Col. of P. & S.; Surg. Cook County and Emergency Hosps.

Fenn, Curtis T. (R) Rush, 1867; 3030 Michigan ave., 8 to 9, 12 to 2, 6 to 7:30; tel. 8143.

Fenner, H. B. Office and res. 583 W. Ohio, 9 to 11, 3 to 5, after 7.

Fernitz, Gustav, (R) Louisville Med. Col., 1881; office and res. 1020 Milwaukee ave., 7 to 9, 2 to 3, 7 to 9; tel. 4202.

Fessel, Christian, (R) Univ. of Berlin, 1826; office and res. 220 Fremont, 8 to 10, 2 to 4.

Fillmore, E. A. (R) Trinity Univ., Toronto, 1884; 395 W. Harrison, 9 to 11, 2 to 4, evenings; tel. 4637.

Finucane, A. D. (R) Bellevue, 1884; 728 31st.

Fisher, John (R) P. & S., 1887; office and res. 452 Wells, till 9, after 5.

Fiske, G. F. (R) Yale, 1883; office, Chicago Opera House blk., 9:30 to 11:30; tel. 1206; res., 438 LaSalle ave., 12 to 1:30; tel. 3027.

Fischer, E. J. (R) Rush, 1880; 557 Sedgwick, 8 a. m. to 9 p. m.; tel. 3607.

Fischer, Gustav (R) Prague Univ., Bohemia, 1871; office, 136 W. 12th st., 9 to 10; 585 Center ave., 10 to 12, 4 to 6, 9 to 10; res. cor. 12th and Johnson; tel. 4378.

Fitch, Calvin M. (R) Univ. of City of N. Y., 1852; office and res. 645 W. Monroe, 1 to 4, Sat. and Sun. excepted; tel. 7041.

Fitch, Henry A. (R) Univ. Mich., 1878; office and res. 884 W. Madison.

Fitch, T. Davis (R) Rush, 1854; office and res. 296 W. Monroe, 10 to 2, 6 to 8; tel. 4504. Consulting Phys. Washingtonian Home; Prof. Emeritus of Gynæcology, Woman's Med. Col.

Fitch, Walter M. (R) Rush, 1885; office and res. 645 W. Monroe, 11 to 12, 6 to 7; tel. 7041.

Fitzgerald, F. W. (R) Rush, 1880; office and res. 318 S. Center ave., 8 to 10, 12 to 2, 6:30 to 8; tel. 4071.

Fitzgerald, J. H. (R) Dublin Univ., 1886; office and res. 492 W. 12th, till 12, after 4.

Flanders, Alicia A. (H) Hahnemann, 1876; office and res. 2708 South Park ave.

Fleming, C. K. (R) Syracuse, 1873; Chi. Med., 1886; office, 3035 South Park ave., 11 to 12, 4 to 5, 7 to 8; res. 3234 Groveland ave.

Fleming, John M. (R) Rush, 1872; office, 112 W. Washington, 9 to 11, 4 to 5, 7 to 9; tel. 4753; res. 113 Warren ave.

Flower, W. Z. (R) P. & S., 1877; 454 W. Madison.

Foote, D. A. (H) Chi. Hom., 1887; Chi. Hom. Hosp.

Formaneck, Fred R (R) Rush, 1887; office, 136 W. 12th, 11 to 12; tel. 4838; res. 585 Center ave., till 10, after 6; tel. 9104.

Forrester, Jessie G. (E) Bennett, 1886; office, cor. Clark and Huron, 2 to 5; res. 1426 Dunning, 10 to 12, 6 to 9.

Foster, A. H (R) P. & S., N. Y.; office, 132 La Salle, 1 to 2, 6:30 to 7:30; tel. 7029; res. 779 W. Monroe.

Foster, B. D. (R) Jefferson, 1883; office and res., 37 Rush, 9 to 12, 2 to 2, 7 to 9; tel. 3316.

Foster, F. H. (H) Hahnemann, 1872; office, 103 State, 9 to 12, 3 to 5; res. 499 Webster ave.

Foster, R. N. (H) Hahnemann, 1869; office and res. 10 Warren ave., 8 to 10, 5 to 7. Prof. Obstetrics, Chi. Hom. Med. Col.

Foulks, C. Allison (R) Chi. Med., 1885; office, 279 State, 4 to 6 p.m.; tel. 320; res. 4700 State, 10 to 12, 6 to 8; tel. 9807.

Frank, J. (R) Univ. Buffaloniensis, 1882; office and res. 17 Lincoln ave., 8 to 9, 12 to 2, 7 to 8.

Franke, Jno. G. (R) Rush, 1868. Chi. Med., 1871; office and res., 890 W. 21st, 8 to 10 2 to 4; tel. 9002.

Fraser, Donald (R) McGill, 1868; Royal P. & S., Edinburg, 1869; office and res. 2 Winthrop pl.

Frazer, C. A. (H) Chi. Hom., 1887; Chi. Hom. Hosp.

Fredigke, Chas. C. (R) P. & S., 1886; office, 3500 State, 10 to 1, 3 to 7; tel. 9875; res. 445 37th, till 10, 1 to 3, 7 to 8.

Freeman, Arthur B. (R) Rush, 1885; Philadelphia Dental Col., 1886; 325 W. Madison, 9 to 12, 1 to 5; tel. 4521. Prof. of Comparative Dental Anatomy, Northwestern Dent. Col.

Freer, Otto T. (R) Rush, 1879; office and res. 153 N. State, 9 to 10; 1 to 2:30, 6 to 7:30; tel. 3555.

French, S. M. (H) Hahnemann, 1882; office and res. 64 23d, 8 to 12, 7 to 9.

Freschonsky, Axel (R) Univ. Copenhagen, 1881; office and res. 44 Beethoven pl.

Freund, A. L. (E) Bennett, 1877; office and res. 322 Wells, till 9, 1 to 3, 6:30 to 7:30; tel. 3367.

Frink, Charlotte E. (H) Office and res., 47½ Pine, 10 to 2.

Fritts, L. C. (H) Chi. Hom., 1888; office and res. 285 W. Monroe, till 10, 5 to 7; tel. 4371.

Froom, Albert E. (R) Chi. Med., 1886; office, 3901 Wentworth ave., 10 to 11:30, 3 to 5, 7:30 to 9; res. 3726 La Salle, till 8:30, 12 to 2, 5 to 7:30; tel. 9869.

Frothingham, Herbert H. (R) Chi. Med., 1885; office, 235 State, 2 to 3; tel. 2349; res. 4306 Lake ave., 9 to 10:30, 4:30 to 6; tel. 9982. Surgeon St. Luke's Hosp. Dispensary; Demonstrator of Anatomy, Chi. Med. Col.

Fuersattel, Chris. Wooster, 1873; office, 347 5th ave., 9 to 11, 2 to 4; res. 781 W. 12th; tel. 7229.

Fuller, Charles Gordon (H) Chi. Hom. Med. and N. Y. Ophthalmic Hosp. Col., 1881; office, 38 and 39 Central Music Hall, 10 to 1, 3 to 4; res. Evanston; tel. 8. Eye and Ear Surgeon, Chi. Ave. Disp. and Bethesda Mission.

G

Galloway, Rhoda M. (R) Wom. Med., 1887; Wom. Hosp., 32d and Rhodes ave.

Gandey, Thos. M. (R) Rush, 1882; office and res., 1593 Milwaukee ave.

Gardiner, Edwin J. (R) San Carlos Med. Col., Madrid, 1878; office, 70 Monroe, 9:30 to 1, 3:30 to 5; res., 85 Astor. Surg. Eye Dept., Ill. Charitable Eye and Ear Infirmary.

Gardiner, Frank H. (H) Chi. Hom. Med. Col., 1882; office, 126 State, 9 to 5; res., Hyde Park. Lecturer on Dental Surgery, Chi. Hom. Med. Col.

Garrott, E. (R) Univ. Maryland, 1856; office and res., 751 Washington boul., 8 to 10, 1 to 2, 6 to 7:30; tel. 7154.

Garvin, W. D. (H) Hahnemann, Phila., 1884; office, 81 N. Clark, 10 to 4; res., 122 Winchester ave., till 10, 4 to 9 p. m.

Gaston, Emma F. (R) Wom. Med., Pa., 1876; office and res., 2308 Michigan ave., 10 to 12, 5 to 7:30; tel. 8184.

Gatchell, Charles (H) Pulte Med. Col., Cin., 1874; office and res., 2954 Prairie ave., 9 to 10, 4 to 5; tel. 8102. Attg. Phys. Cook Co. Hosp.

Gates, James H. (R) Rush, 1884; office and res., 470 W. Adams.

Geiger, H. (R) Heidelberg, 1861; office and res., 481 N. Clark, till 9, 1 to 2:30, after 6; tel. 3027.

German, Wm. H. (R) Michigan Col., 1883; office, 163 State, 11:30 to 2; res., Morgan Park, 8 to 9, 5 to 7:30.

Gfroerer, George S. (R) P. & S., 1885; cor. Halsted and Polk; tel. 4388.

Gibbs, Lucius O. (E) Bennett, 1871; office, 181 S. Clark, 8 a. m. to 8 p. m.; tel. 1339; res., 1043 W. Adams.

Gibson, Charles B. (R) Office, 81 Clark, 8:30 to 5; res. 113 Hoyne ave., Prof. of Inorgan. Chem., Col. of P. & S.; Prof. of Chem. & Metallurgy, Chi. Col. Dental Surg.

Giegerich, C. (E) Bennett, 1881; office and res., 449 Milwaukee ave., 7 to 9, 1 to to 3, after 7; tel. 4575.

Giljohann, Emil (R) Rush, 1886; office and res., 603 31st, 9 to 11, 1 to 3, 7 to 8.

Gill, M. (R) Vermont Academy of Med., 1835; office, 171 N. Clark; res., 208 N. Clark, 10 to 1, 3 to 6.

Gilman, John E. (H) Hahnemann, 1871; office, Central Music Hall, 1 to 3, except Sun.; res., 455 Washington boul., 8 to 9, 7 to 9; tel. 7298. Prof. of Physiology, Histology and Sanitary Science, Hahnemann Med. Col.

Gilmore, Arnold P. (R) Jefferson, 1874; office, 70 Monroe, 10 to 2; tel. 80; res., 113 Cass. Phys. Home for Incurables; Asst. Surg. Eye Dept. Ill. Charitable E. & E. Infirmary.

Goldspohn, Albert (R) Rush, 1878; office and res., 163 Lincoln ave., 9 to 10, 7 to 8; tel. 3634. Mem. German Hosp. Staff.

Goodner, G. W. (R) Chi. Med. 1869; office and res., 2204 Archer ave., 9 to 10, 2 to 3:30, evenings; tel. 8204.

Gordon, A. H. (H) Hahnemann, 1887; office and res., 207 E. Chicago ave., 8 to 9, 1 to 3, 7 to 8; tel. 3274.

Gore, J. R. (R) Univ. of N. Y. City, 1850; office, 187 LaSalle, 12 to 2; res., 2606 Prairie ave., before 11, after 3; tel. 8022 and 2500.

Gradle, Henry (R) Chi. Med., 1874; office, Central Music Hall, 9:30 to 3; tel. 5558; res., Waukegan. Oculist, Michael Reese Hosp.

Graham, C. (E) Bennett, 1876; 363 State, 10 to 3, 6 to 8; Sun., 10 to 12.

Graham, David W. (R) Bellevue, 1872; office, 133 Clark, 11 to 1; tel. 1339; res., 105 Warren ave., 8 to 9, 5:30 to 7:30; tel. 7026. Prof. of Surgery, Woman's Med. Col.; Prof. Surgical Emergencies, Chi. Training School; Attg. Surg. Presbyterian Hosp. and Cent. Free Disp.

Graham, James F. (R) Univ. of N. Y., 1880; office and res., 278 Bissell, 8 to 11, 2 to 5, nights.

Gray, Albert S. (E) Bennett Med. Col., 1887; 103 State, 11 to 4.

Gray, Allen W. (R) Chi. Med., 1868; office, 722 W. Lake, 10 to 12, 3 to 5; res., 667 Fulton.

Gray, Ethan A. (R) Rush, 1887; 667 Fulton.

Gray, J. Lucius (R) Chi. Med., 1885; office, 70 E. Monroe; 12 to 2; tel. 80; at Co. Physician's Office before 9, after 5; tel, 3655. Supt. Detention Hosp. Insane; Attg. Phys. Nervous Diseases, S. Side Free Disp.

Gray, John T. (R) P. & S., N. Y., 1879; office, 181 W. Madson, 2 to 5, 7 to 9; tel. 4152; res., 223 Colorado ave., till 1 p. m.; tel. 7007.

Graves, Kate I. (H) Hahnemann, 1885; office and res., 3629 Vincennes ave., 8 to 11, 5 to 8.

Green, Isadore L. (H) Hahnemann, 1886; office and res., 315 Lincoln ave., 9 to 12, 7 to 8; tel. 3317.

Greenfield, C. E. (R) Rush, 1886; office, 378 Blue Island ave., 11 to 12; res., 260 S. Halsted.

Greensfelder, Louis A. (R) Chi. Med., 1887; Michael Reese Hosp.; tel. 8212. Interne Michael Reese Hosp.

Greer, J. H. (E) Bennett, 1875; office and res. 109 Loomis.

Greer, Robert (E) Office, 127 La Salle, 8 a. m. to 8 p. m.; Sun., 10 to 12; tel. 1339; res. 307 S. Oakley ave.

Griffin, Byron Wilson, (R) Rush, 1876; 937 W. Madison, till 9, 2 to 3, 6 to 7; tel. 7195. Lecturer on Diseases of Nervous System, Chi. Training School for Missionaries.

Griggs, L. P. Office and res. 322 Mohawk.

Griswold, W. R. (R) Albany, 1853; 22 Park ave., 8 to 9, 1 to 2:30, 7 to 8; tel. 7018.

Gross, James E. (H) Hahnemann, Phila., 1850; office, 48 Madison; tel. 2587; res. Hotel Woodruff.

Gross, M. M. (H) Cleveland Hom., 1858; office and res. Hotel Woodruff, 11 to 1; tel. 8179.

Grosvenor, Lemuel Conant, (H) Western Hom. Med. Col., Cleveland, 1864; office and res., 185 Lincoln ave., 8 to 10, 3 to 5; tel. 3404. Prof. of Sanitary Science in Chi. Homeo. Col.

Guerin, J. (R) Rush, 1866; office, 3102 State, 10 to 12; res. 4440 Michigan ave.

Guffin, E. L. (E) Bennett, 1884; office, 125 State, 1 to 4; res. Oak Park.

Gwynne, E. E. (H) Hahnemann, 1879; 407 Center, 8 to 10, 1 to 2, 5 to 7.

H

Haerther, A. G. (R) Chi. Med., 1883; office, 64 State, 2 to 6; tel. 2558; res., 696 W. Lake.

Hahn, H. S. (R) 70 State, 10 to 12, 4 to 5; tel. 1336; res. 330 Warren ave.

Haight, Vincent, (R) 1001 Madison, till 10, 4 to 8. Prof. of Anatomy, Am. Den. Col., Wed. and Sat., 9 to 10; tel. 7057.

Haines, Walter S. (R) Chi. Med., 1873; office, Laboratory of Rush Med. Col.; res. Waukegan. Prof. of Chemistry, Pharmacy and Toxicology, Rush Med. College.

Halbert, H. V. (H) Hahnemann, 1887; 2400 Prairie ave., 8 to 9, 1 to 2, 5 to 6; tel. 8182. Adj. Chair of Anatomy, Hahnemann.

Hale, Albert B. (R) Chi. Med. Col., 1886; 65 22nd; tel. 8286.

Hale, Edwin M. (H) Cleveland Hom. Col., 1859; 65 22nd, 8 to 9, 1 to 3, 7 to 8; tel. 8286.

Hall, George A. (H) Homeo. Med. Col., Phila., 1856; office and res. 2400 Prairie ave., 9 to 12, 7 to 8; tel. 8182. Surgeon Hahnemann Hosp.

Hall, Geo. Cleveland, (E) Bennett. 1888; office, 533 State, 9 to 11, 3 to 5; tel. 667; res. 3139 Butterfield, 7 to 9 p. m.; tel. 8330.

Hall, Junius M. (R) P. and S. N. Y., 1874; office, McVicker's Theatre building, 12 to 2; res. 2 Washington pl.

Hall, Randolph N. (R) Rush, 1882; office, 210 W. Indiana, 10 to 11, 5 to 6; tel. 4562; res. 676 W. Indiana, 11:30 to 1; Sun., 12:30 to 1:30; tel. 7064.

Hall, R. S. (R) Rush, 1872; 863 W. Harrison till 9, 4 to 5; tel. 7124. Clinical Lecturer on Midwifery and in charge Outside Obsteric Dept., Wom. Med. College.

Hall, William E. (R) Rush, 1878; office and res., 3504 State, 2 to 4, 7 to 9.

Hallowell, Clement H. (H) Boston Univ. School of Med., 1879; 3240 Cottage Grove ave., 2 to 4, 7 to 9; tel. 8354.

Hammon, Glenn M. (R) Rush, 1881; office, 683 W. Adams, till 8, 12 to 2; tel. 7216.

Hammond, J. D. (R) Rush, 1884; 240 Wabash ave., 8 a. m. to 7 p. m.; tel. 2810; res. Leland Hotel.

Hancock, Joseph L. (R) Chi. Med. Col., 1883; 3120 Vernon ave., 9 to 11, 7 to 8; tel. 8494.

Hanley, Wm. (R) Office, 193 Clark, 12 to 4; tel. 1339; res., 600 VanBuren, 7 to 9, 6 to 8; tel. 7119.

Hanna, E. A. (R) Chi. Med., 1886; 3857 Vincennes ave., 11 to 12, 4 to 5, 7 to 8; tel. 9843.

Hannah, Helen M. (H) Chi. Hom., 1881; office and res. 3030 Prairie ave., 8 to 10, 2 to 4; tel. 8108.

Hanson, Z. P. (R) Rush, 1861; office and res. 306 Washington blvd., 7:30 to 9:30, 2 to 4, 6:30 to 7:30.

Harcourt, Luke Arthur (R) Buffalo Med. Col., 1868; office and res. 1214 W. Jackson, 8 to 9, 1 to 2, 7 to 8; tel. 7206.

Harper, John E. (R) Univ. of N. Y., 1878; office, 163 State, 9 to 1; tel. 5480; 607 W. Van Buren, 4 to 6; res. 148 Loomis. Prof. of Ophthalmology and Otology, Col. of P. & S.; Chief of Eye and Ear Dept., W. Side Free Disp.; Oculist at Ill. Charitable E. and E. Infirmary.

Harris, D. J. (R) Univ. of Mich., 1864; office and res. 3145 Vernon ave.

Harris, M. L. (R) Rush, 1882; office and res. 482 Milwaukee ave., 9 to 10, 1 to 3, 7 to 8, night; tel. 4029.

Harrison, Wallace K. (R) Yale Col. Med., 1874; P. & S., Chi., 1882; office, 103 State, 11 to 3; tel. 1339; res. 295 N. State, till 9, 6 to 8; tel. 3537. Prof. Medical Chemistry, Col. P. & S.

Hart, M. G. (E) Bennett, 1883; office, 511 State, 10 to 12, 2 to 4, 7 to 9; tel. 667; res. 1558 Wabash ave., before 10, and 5 to 7; tel. 8247. Prof. of Physiology, Bennett Col.

SPRING·BANK.

... TO THE PUBLIC. ...

THE view on inside front cover represents the grounds and residence of the late CAPT. T. L. PARKER, situated about three miles from the city of Oconomowoc, on a lake bearing the same name. Having secured and fitted this beautiful place as a summer resort, I am prepared to give tourists every accommodation. Having a large farm in connection with the grounds, I have my own dairy products, poultry yards, gardens and a fine livery. The fishing is fine, as the lakes abound in black bass, pickerel, etc. For accessibility, position, beauty of location, salubrity of atmosphere, and the purity of water from its numerous springs, it has no equal in Wisconsin. Parties seeking health or pleasure will be well repaid a visit to this Eden of the Northwest. The cuisine will be of the best the markets afford. Guests bringing their own conveyances and coachmen will find ample accommodation for them at

"SPRING BANK."

Special rates to season guests, for which address the proprietress,

MRS. J. A. BURTIS.

Lock Box 464. Oconomowoc, Wis.

Open from May 15th to October 15th.

Hartman, F. S. (R) Rush, 1885; office and res. 262 S. Halsted.
Harvey, W. S. (H) Hahnemann Med. Col.; office and res. 565 W. Madison, 8 to 9, 1 to 3, 7 to 8; Sundays, 9 to 10, 2 to 4; tel. 7172.
Hasbrouck, R. A. (E) Bennett, 1882; 511 State, 8 to 11, 2 to 4, 7 to 8.
Hasse, F. (E) Bennett, 1883; office and res. 570 Milwaukee ave., till 10, 2 to 4.
Hatfield, Marcus P. (R) Chi. Med., 1872; office, 3235 S. Park ave., 8:30 to 9:30, 2 to 4; tel. 8544; res. 3301 Forest ave.; tel. 8161. Prof. of Diseases of Children, Chi. Med. Col.; Lecturer Ill. Col. of Pharmacy; Phys. Children's Dept. S. Side Free Disp and Chi. Orphan Asylum.
Hattermann, Carl (R) Rush, 1886; office and res. 994 Milwaukee ave., till 10, 12 to 3, 6 to 8.
Haven, Joseph (R) Rush, 1880; office and res. 90 Warren ave., 12 to 2; tel. 7225. Dean, N. W. Col. Dental. Surg.
Hawkes, William J. (H) Hom. Col., Phila., 1867; office, Central Music Hall, 1 to 4; tel. 2642; res. 241 Dearborn ave., till 9:30, 6 to 7:30; tel. 3037. Prof. of Materia Medica and Therapeutics, Hahnemann Med. Col.
Hawley, George F. (R) P. & S., N. Y., 1867; office, 125 State, 12 to 5; tel. 5442; res. 176 S. Ashland ave. Asst. Surg. Ear Dept., Ill. Charitable E. and E. Infirmary; Surgeon, Throat and Chest Hosp.
Hay, Walter (R) Columbia Univ., 1853; office, 243 State, 8 to 12, res. Hyde Park. Prof. of Nervous and Mental Diseases and Med. Jurisp., Chi. Med. Col.
Hayes, Edwin H. (H) Hahnemann Med. Col., 1884; office and res. 243 State, 10 to 5; tel. 5349.
Hayes, Justin (R) Cleveland Med. Col., 1850; office, 240 Wabash ave., 10 to 4; tel. 5457; res. Western Springs.
Hayes, Plymmon (R) Rush, 1872; office, 240 Wabash ave., 11 to 2; res. 159 46th, till 8, 6 to 7:30; tel. 9984. Prof. of Gynæcology and Electro-Therapeutics, Chi. Policlinic, Med. Dept. South Side Free Dispensary.
Hayman, L. B. (R) Rush, 1886; office, 71 Dearborn, 9 to 11; tel. 5453; res. 372 W. Monroe, till 8:30, 4 to 8; tel. 4736. Med. Dept. St. Luke's Hosp. Disp.
Hayman, William Henry (R) Rush, 1886; 530 W. Indiana, till 10, 12 to 2, 5 to 7; tel. 7073.
Hayner, Jennie E. (R) Woman's Med., 1880; office, Central Music Hall, 1 to 2; tel. 2642; res. 192 W. Erie, 10 to 12, 5 to 6; tel. 4612.
Haynes, L. B. (R) Rush, 1886; 71 Dearborn, 9 to 11; tel. 2453; res. 372 W. Monroe, till 8:30, 4 to 8; tel. 4736.
Haynes, C. M. (H) Homeo. Hosp. Col., Cleveland, 1877; office and res. 3838 Dearborn, 10 to 12.

Haynes, G. F. (E) Eclec. Med. Col. of Cin., 1871; office and res., 3838 Dearborn, 4 to 6.
Hays, J. R. (R) Vienna, 1863; 642 W. 21st.
Heckenbach, J. A. (R and E) Bennett, 1882; office and res., 657 N. Halsted, 8 to 10, 2 to 4.
Hedges, Samuel Parker (H) Hahnemann, 1867; office, Central Music Hall, 11 to 4; tel. 2642; res. cor. Evanston and Graceland aves., Lake View, 7 to 8. Con. Phys., Chi. Nursery and Half-Orphan Asylum.
Heegard, B. L. (H) Chi. Hom. Med., 1880; 174 Eugenie.
Heffernan James (R) Rush, 1881; 106 State, 9 to 11, 2 to 4, 7 to 8.
Heise, Ellen H. (R) Wom. Med., 1887; 1010 W. Harrison, 8 to 9, 6 to 7; tel. 7201; Oak Park, Wed. and Sat., 2.30 to 4. Asst. Chair of Anatomy, Wom. Med. Col.
Helm, Scott (R) Rush, 1883; 204 31st, 1 to 2, 5 to 8; tel. 8042.
Helmuth, Carl A. (R) Univ. of Berlin, 1844; office, 26 N. Clark, 3:30 to 5; res. 624 Lincoln ave., 8 to 10, 1 to 3.
Hemmi, Stephen A. (E) Bennett, 1884; office, 567 W. Chicago ave., 7 to 9, 1 to 3, 7 to 9; cor. Wood and Division, 10 to 11, 4 to 5; tel. 7093.
Hempstead, C. W. (R) Univ. of Penn., 1852; office and res. 672 W. Monroe, till 9, 1 to 2; tel. 7023.
Henderson, N. H. (R) P. & S., 1886; office, 516 Opera House bldg., 12 to 2; res. Englewood.
Hendriks, J. W. (R) Office, 159 North ave., 9 to 10, 1 to 2, 7 to 8; res. 232 North ave., till 8, 12 to 1, 6 to 7; tel. 3340.
Henrotin, Ferdinand (R) Rush, 1868; office, Opera House bldg., 1 to 2:30; tel. 3062; res. 353 La Salle ave., 7 to 8, 12 to 12:30, 6:30 to 7:30; Sun., 8 to 10.
Henry, F. H. (R) Chi. Med., 1886; office, 631 Center ave., 11 to 12, 4 to 7; tel. 9137; 806 S. Ashland, 9 to 11; res., 240 W. 14th; tel. 9030.
Hequembourg, J. E. (R) Rush, 1882; office, Cent. Music Hall, 11 to 1; res, 513 Fullerton ave., till 9, 7 to 8; tel. 3432.
Herrick, James B. (R) Rush, 1888; Cook Co. Hosp.
Herz, Karl (R) Karl Ferdinand Univ., Prague, Austria, 1884; 448 N. Clark, 11 to 12, 4 to 6; tel. 3312.
Hess, F. A. (R) Rush, 1873; 247 Division.
Hessert, G. (R) Wurzburg, 1864; 267 Erie, 1 to 2; tel. 3054.
Heuchling, Th. (R) Wurzburg, 1864; office, 218 La Salle, 2 to 3; tel. 5860; res. 509 N. Clark, 7 to 9, 12 to 1, 6 to 8; tel. 3027.
Hiatt, Alfred H. (R) Ohio Med. Col., 1846; office, 40 Cent. Music Hall, 11 to 4; tel. 2503; res. Wheaton, Ill.
Hickey, Rachel (R) Wom. Med., 1887; Cook Co. Hosp. Asst. Chair Anatomy, Wom. Med. Col.
Hickox, K. L. (H) Hahnemann, 1887; 358 37th, 8 to 10, 4 to 6; tel. 9881.
Higgins, Charles C. (R) Univ. of Iowa, 1863; office, Opera House bldg., 11 to 4; Sun., 12 to 2; tel. 1339; res. 324 Wells, 7 to 9; tel. 3367.

Hildebrand, H. E. (E) Bennett, 1877; office, 426 State, 8 to 10, 2 to 5, 7 to 9; tel. 306; res. 2407 Dearborn.

Hillegas, W. R. (R) Albany, 1881; office and res. 3600 Vernon ave., 8 to 10, 3 to 5, 7 to 8.

Hinde, Alfred (R) Rush, 1878; 16 Laflin, till 8, 4 to 7.

Hinish, W. W. (R) Miami, 1875; office, 155 Washington; res. 229 Wilmot ave.

Hinkle, Abbie A. (H) Hahnemann 1887; office and res., 996 Millard ave., 8 to 10:30, 5 to 8.

Hinz, Augusta (R) Kiel, Germany, 1866; office and res. 399 Division, 8 to 9, 2 to 5; tel. 3152.

Hoadley, Albert E. (R) Chi. Med., 1872; office and res. 683 Washington blvd., till 8, 11 to 12, 5 to 7; tel. 7020. Surgeon and Prof. of Anatomy, Col. P. & S., Chi.; Surg. Cook Co. Hosp. and Chi. Policlinic Hosp.; Prof. of Surgery, Chi. Policlinic.

Hoag, J. C. (R) Chi. Med., 1882; office and res, 305 E. Division, till 10, 6:30 to 7:30, eve.; tel, 3284. Demonstr. Histol., Chi. Med. Col.

Hobart, Henry M. (H) Hahnemann, 1876; office and res. 402 Center, 8 to 9, 1 to 2, 6 to 7; tel. 8109. Attg. Phys. at Chi. Nursery and Half-Orphan Asylum; Prof. Materia Medica, Chi. Hom. Med. Col.

Hobbs, John O. (R) Chi. Med., 1872; office, 455 Center ave., 8 to 9, 1 to 2, 6 to 7; tel. 9081; res. 432 W. Jackson; tel. 4170.

Hodson, F. A. (R) Wooster Univ., 1882; office and res. cor. Van Buren and Throop, till 9, 4 to 8, night; tel. 4188.

Hoegelsberger, Franz C. (R) Univ. of Vienna, 1872; office and res. 297 Clybourn ave., 7 to 9, 1 to 3, 7 to 9.

Hoegelsberger, Hans (R) Univ. of N. Y. City, 1884; office and res. 297 Clybourn ave., 9 to 10, 5 to 7.

Hoelscher, J. H. (R) Chi. Med., 1885; office and res. 451 North ave., till 9, 2 to 4, 7 to 9; tel. 3206.

Hogan, Cornelius (H) Hahnemann, 1881; 126 S. Halsted, 10 to 12, 3 to 5, 7 to 8:30; tel. 4068.

Hogan, Sarah J. (E) Bennett, 1886; 279 N. May.

Hollister, Elmer L. (R) Chi. Med., 1882; office, 171 22d, 9 to 11, 1 to 4, 7 to 8:30; tel. 8165; res. 2123 Michigan ave.

Hollister, John H. (R) Berkshire Med. Col., 1847; office, 70 Monroe, 12 to 3; tel. 5480; res. 3440 Rhodes ave., till 8, after 6; tel. 9827. Prof. of Clinical Med., Chi. Med. Col.; Physician, Mercy Hospital.

Holmboe, A. (R) Col. P. & S., Chi., 1886; 196 Wells; tel. 3664. Asst. Surgeon, Emergency Hospital.

Holmes, Bayard (H) Chi. Hom. Med., 1885; office, 125 State, 1 to 2; tel. 9910; res. 1535 Bowen ave., till 10, after 4.

Holmes, Ed. L. (R) Harvard, 1854; office, 112 Clark, 10 to 1; res. 530 W. Adams. Prof. of Diseases of Eye and Ear, Rush Med. Col.; Attg. Oculist and Aurist, Presbyterian Hosp.; Senior Surg., Ill. Charitable E. and E. Infirmary.

Holmes, Saul J. (R) Rush, 1876: N. Y. Med. Univ., 1877; office and res. 315 Fulton, 1:30 to 2:30, 7 to 9; tel. 4193. Lecturer on Pathological Anatomy and Pathological Histology, Rush Med. Col.

Holroyd, Eugene E. (R) Col. P. & S., Keokuk, 1878; office and res. 412 Park ave., till 9, 1 to 3, 6 to 7:30; tel. 7106.

Hood, C. T. (R & H) Col. P. & S., 1884; office and res. 255 Western ave., 8 to 9, 1 to 2, 6:30 to 7:30; tel. 7126 and 7249.

Hooper, Henry (R) Harvard, 1869; office and res. 215 Dearborn ave., till 8, 12 to 1, 5 to 7; at Alexian Hosp. 8 to 9 a. m.; tel. 3171.

Hopkins, J. F. (R) Chi. Med., 1860; office and res. 171 Park ave., till 9, 12 to 2, after 6; tel. 7002.

Horn, F. S. (E) 125 S. Western ave.

Hosmer, Arthur Burley (R) Chi. Med., 1875; office, 168 Dearborn, 8 to 8:30; 6:30 to 7:30; tel. 3330; res. 70 Monroe, 1:30 to 3; tel. 80.

Hottenroth, A. (R) Leipsic, 1859; 498 Wells; tel. 3143.

Hotz, Ferdinand Carl (R) Heidelberg Univ., 1865; office, 103 State, 9 to 1; res. 109 S. Morgan. Attg. Surgeon, Ill. Charitable E. and E. Infirmary.

Howard, William A. (R) Rush, 1882; office, 208 S. Halsted, 10 to 11, 1 to 2, 7 to 9; tel. 4092; res. 977 W. Polk, till 9, 12, 4 to 6; tel. 7152.

Howe, E. J. (R) Univ. of Mich., 1886; office, 336 W. 12th, 2 to 6; tel. 4061; res. 529 13th place, till 10; tel. 7162.

Howe, Julia R. (E) Bennett, 1886; office and res. 257 W. Madison, 9 to 11, 1 to 3.

Howe, O. B. (E) Bennett, 1878; office and res. 239 W. Chicago ave., 8 to 10, 1 to 3, 6 to 9.

Howe, S. W. (R) Hanover, N. H., 1842; 255 W. Madison, 10 to 12, 1 to 5.

Hoyne, Temple S. (H) Bellevue, 1865; office and res. 1833 Indiana ave., 8 to 9:30, 1 to 2, 5 to 7; tel. 8570. Prof. of Skin and Venereal Diseases, Hahnemann Hosp. Col.

Hueston, D. P. (R) P. & S., '88; 339 W. Harrison, 8 to 9, 3 to 8, tel. 4560

Huff, O. N. (R) Univ. of Mich., 1878; office and res. 5 Washington pl., till 9, 12 to 2, 6 to 7; tel. 3601.

Huffaker, T. S. (H) Hahnemann, 1884; office, 3900 Cottage Grove, 9 to 10, 2 to 4, 7 to 8; tel. 9817; res. 245 E. 43d, till 8:45, 11 to 1, 5 to 6; tel. 9913.

Hunt, D. W. (R) Albany, 1856; 11 Clarkson ct.

Hunt, Geo C. (R) Rush, 1884; 410 La Salle ave.; tel. 3312.

Hunt, H. M. (R) P. & S., St. Louis, 1887; office and res. 348 Michigan ave.

Hunt, J. S. (R) Miami, 1855; Jefferson, 1856; 878 Fulton.

Hunt, William Carleton (R) Rush, 1852; office, 41 Clark, 2 to 4; tel. 5453; res. 384 N. State, 8 to 9, 7 to 8; tel. 3324.

Hunter, E. W. (R) Rush, 1877; office, 103 State, 10 to 2; tel. 642; res. 2228 Prairie ave., 8:30 to 9:30, 4 to 5, 7 to 8; tel. 8332.

Hunter, Robert (?) Univ. of N. Y. City, 1846; 103 State, 9 to 4.

Hurlbut, Horatio N. (R) Starling, O., 1840; 2332 Prairie ave.

Hurlbut, V. L. (R) Rush, 1852: office and res., 204 Dearborn, 8 to 10, 2 to 4; tel. 1897.

Hurlbut, John E. (R) Wooster, 1861; office, 108 Washington; res., 138 Park ave.

Hurst, N. N. (R) Jefferson, 1873; 3906 State, 8 to 10, 1 to 2, 8 to 9; tel. 9853; res. 5100 Wentworth ave., 4 to 5; tel. 35, (Englewood Ex.)

Hutchins, A. V. (H) Hahnemann, 1883; 358 Western ave., till 10, 1 to 2, 6 to 8; tel. 7006.

Hutchins, John W. (R) Harvard, 1858; 125 State, 3 to 5; res. 384 W. Adams, till 10, after 6; tel. 4169; Attg. Phys., Cent. Free Disp.

Hutchinson, J. M. (R) Chi. Med., 1867; 173 Blue Island ave., 9 to 10, 2 to 3; tel. 7031; res. 547 W. Jackson, 6 to 7.

Hutchinson, Mahlon (R) Bellevue, 1881; 103 State, 11 to 4; res. Edgewater; tel. 5442.

Hyde, James N. (R) Univ. of Penn., 1869; 240 Wabash ave., 10 to 2; res. 2409 Michigan ave. Prof. of Skin and Venereal Diseases, Rush Med. Col.: Derm. Presbyterian Hosp.; Consulting Phys., Chi. Hosp. for Women and Children; Med. Staff, Michael Reese Hosp.

Hyde, Maranda (E) Bennett, 1878; 259 W. Madison, 9 to 4; tel. 4310 and 4521.

I

Illingsworth, G. M. (R) Chi. Med., 1874; 159 North ave., 8 to 10, 12 to 1, 6 to 8; tel. 3340; res. 150 North ave., 1 to 2.

Ingalls, Frank M. (R) Rush, 1888; 966 W. Lake, at 9, 1 to 2, 7 to 8; tel. 7142; res. 805 W. Monroe, till 9, 12 to 1, 6 to 7.

Ingalls, E. Fletcher (R) Rush, 1871; 70 State, 10 to 2; res. 507 W. Adams until 8, after 7; tel. 7036. Prof. of Laryngology, Rush Med. Col.; Med. Staff, (diseases of the throat and chest) St. Joseph's Hosp.; Attg. Phys., Cent. Free Disp.; Prof. of Diseases of Chest and Throat, Women's Med. Col.

Ingalls, Ephraim (R) Rush, 1847; 34 Throop, 8 to 9, 1, to 3; tel. 4077.

Ingraham, Sereno W. (E) Bennett, 1877; 207 Clark, 11 to 4; tel. 1399; res. 721 Washington boul., 7 to 8, 6 to 7:30; tel. 7175. Phys., Newsboys' Home.

Irwin, John Louis, (R) McGill, 1879; 200 E. Erie, 8 to 9; 1 to 3, 7 to 9; tel. 3045; House Phys. St. Vincent's Infant Asylum.

Irwin, M. F. (R) 389 W. Randolph, 10 to 12, 1 to 4.

Isham, George S (R) Chi. Med. 1884; 70 E. Monroe, 10 to 12; res. 321 Dearborn ave., 2 to 4; tel. 3172; Clinical Asst. to Prof. Surgery, Chi. Med. Col.

Isham, Ralph N. (R) Univ. of N. Y., 1854; 70 Monroe, 12 to 2; res. 321 Dearborn ave., until 9, after 6; tel. 3172. Prof. of the Principles and Pr. of Surg. and Clin. Surg. Chic. Med. Col,; Con. Surg. Presbyterian Hosp.; Attg. Surg. Emergency Hosp. and North Star Disp.

Ives, Franklin B. (R) Rush, 1850; 125 State, 1 to 4; tel. 5442; res. 3226 Calumet ave.

J

Jackson, A. Reeves, (R) Med. Dept. Penn., 1848; 271 Michigan ave., 10 to 12; tel. 630. Pres. and Prof. Gynecology, Col. P. & S.; Chief Gynæcological Dept., W. Side Free Disp.

Jackson, W. M. (R) Chi. Med. 1879; 871 W. 22d, 10 to 12, 3 to 5.

Jacobs, T. K. (R) Med. Col., Ohio, 1880; 1801 State, 10 to 12, 3 to 5, 7 to 9; tel. 8202; res. 189 25th, till 8, 12 to 1, 5 to 7; tel. 8606.

Jacobsen, Harold (R) Kiel (Holstein) Germany, 1881; 2622 Cottage Grove. 8 to 9, 11 to 1, 5 to 7.

Jacobson, Sigismund D. (R) Copenhagen, 1862; 257 Milwaukee ave., 8 to 9, 1 to 3, nights, tel. 4011. Prof. Surgery, Chi. Polyclinic; Consulting Surg., German Hosp.

Jaggard, William Wright (R) Univ. Penn., 1880; 2330 Indiana ave., till 10, 12 to 2, 6 to 7; tel. 8457. Prof. Obstetrics, Chi. Med. Col.; Obstetrician, Mercy Hosp.; Gynæcologist S. Side Dispensary.

Jaques, W. K. (R) Chi. Med., 1887; 3102 State, 11 to 12, 3 to 4, 7 to 8; tel. 8166.

Jay, Milton (E) Ec. Med. Inst., Cincinnati, 1859; N. W. cor. State and Madison, 2 to 5; tel. 624; res. 2510 Indiana ave.; tel. 8136. Dean and Prof. of Surgery, Bennett Col., Pres. Nat. Eclectic Med. Ass'n.

Jennings, P. G. (R) Laval Univ., 1876; 3452 Halsted, 1 to 3 and after 7; tel. 9589.

Jensen, P. C. (R) Univ. of Mich., 1882; Chicago Opera House bldg., 10 to 12, 1 to 4, 7 to 8, (except Sun. and Wed.); tel. 1206.

Jerome, Levi R. (R) Diploma Lewis Co. Med. Soc., 1844; 152 La Salle; room 13, 12:30 to 3; res. Stone ave., LaGrange, till 9, after 4.

Jesperson, Thomas (R) Rush. 1888; 1606 Milwaukee ave., 8 to 10, 4 to 7.

Jirka, Frank J. (R) Rush, 1880; 631 Centre ave., 9 to 10:30; 2 to 3; cor. 12th and Jefferson, 11 to 12, 4 to 5; tel. 9157; res. 804 S. Ashland ave., tel. 9137.

Johnson, Claes William (R) Rush, 1880; 111 E. Chicago ave., till 9, 12 to 2, 6 to 8; tel. 3214. House Surg, St. Joseph's Hosp.; Phys. and Surg. Swedish Home, Mercy.

Johnson, Frank S. (R) Chi. Med., 1881; 4 16th, 9 to 12, 6 to 7; tel. 8239. Prof. Gen. Pathology and Path, Anatomy, Chi. Med. Col.

Johnson, H. A. (R) Rush, 1852; 4 16th, 9 to 1; tel. 8239. Emeritus Prof. Prin. and Pr. Med. and Clin. Med., Chi. Med. Col., Consulting Physician Mercy Hospital.

Johnson, Joseph H. S. (H) P. & S., 1883; Hahnemann, 1884; 915 W. North ave., till 9, 12 to 1, after 6; tel. 4624.

Johnson, W. J. (R) Chic. Med., 1868; Chic. Col. Dental Surg., 1885; 300 Thirty-first; tel. 8145.

Johnstone, Stuart, (R) P. & S., Chi., 1886; 134 S. Halsted, 9 to 10, 2 to 3, 7 to 8; tel. 4133; res. 595 W. Monroe, till 8:30, 1 to 1:30, after 6:30; tel. 7205.

Jones, R. W. (R) P. & S., 1884; cor. Halsted and Washington, 3 to 5; tel. 4031; res. 310 Loomis, 8 to 10, 1 to 2, 7 to 8; tel. 7110. Clinical Instructor N. Side Disp.

Jones, Samuel J. LL.D., Univ. of Penn., 1860; Argyle bldg.; 9:30 to 1, 4 to 5; tel. 2466; res. 487 Dearborn ave.; tel. 3014. Prof. Ophthalmology and Otology, Chi Med. Col.; Oculist and Aurist, St. Luke's Free Hosp.

Joy, H. W. (R) Chi. Med., 1882; cor. State and Harrison, 11:30 to 12:30, 4 to 5, 8 to 9, Sun., 3 to 5; tel. 22; cor. Clark and Harrison, 10 to 11, 7 to 8; tel. 1401; res. 1355 Wabash ave.

Julson, J. Univ. Christiani, Norway, 1849; 246 Milwaukee ave., 8 to 10, 1 to 2, 5 to 8; tel. 4011.

K

Kadison, A. P. (R) Munich, 1874; 503 S. Canal, till 10, 5:30 to 7:30.

Kales, John D. (R) Harvard, 1887; 353 N. Clark, 9 to 12, 3 to 5:30; Res. 291 Huron, after 6 P. M; Tel. 3125.

Kalkstein, L. V. (R) Berlin, 1848; 252 Orchard, till 9, 3 to 6.

Karst, F. A. (H) Hahnemann, 1877; 521 Wells, 9 to 11, 2 to 4, Tues., Wed. and Fri., 7 to 9 P. M.

Kauffman, A. E. (R) Rush, 1885; 325 W. Madison, 11 to 12, 4 to 6, 7 to 8; Sun., 12 to 1; Tel. 4521. Demonstrator Chemistry, Rush Med. Col.

Kean, John, (E) 167 Clark; res, 394 Washington Blvd.

Kearsly, Mary J. (R) Wom. Hosp. Med., 1888; 2834 Main, till 9, 1 to 3, after 7; tel. 7195.

Keeler, Horatio (H) Hahnemann, 1872; 163 State, 1 to 3; res. 3543 Prairie ave., till 9:30, 6 to 7.

Kelleher, Michael W. (R) Bellevue, 1885; 287 W. 12th, 8 to 10:30, 2 to 4.

Kellner, M. G. (R) Rush, 1882; 358 Larrabee, till 9, 1 to 2, 6 to 9.

Kemp, N. C. (H) Hahnemann Med. Col., Chic., 1886; 3904 Indiana ave., 8 to 10, 1 to 3, 6:30 to 7:30; Sunday, 1 to 3; tel 9941.

Kennedy, J. E. (R) Trinity Col., Toronto, 1861; Univ. Toronto, 1866; 262 Walnut.

Kenning, R. H. (R) P. and S., Manitoba, 1883; 1333 W. Lake, 8 to 10; 3 to 4; tel. 7217.

Kerler, Chas. L. (R) Vienna, 1854; 509 W. 12th.

Kerlin, E. Iles, (R) Univ. of Penn., 1886; 159 North ave., 2 to 3; tel, 3340; res. 1102 N. Halsted, till 9; 11 to 1, 5 to 7 and nights; tel. 3966.

Kernahan, G. (R) Rush, 1880; cor. Randolph and Halsted, 10 to 12, 2 to 5; night.

Kerr, E. E. (R) Chi. Med, 1887; cor. 26th and Calumet ave., 9 to 10, 12:30 to 2; after 7; tel. 8269.

Kewley, J. R. (R) Chi. Med., 1886; 1700 Wabash ave., 9 to 11, 3:30 to 5; tel. 8319; res. 3111 Wabash ave, till 8:30, 1 to 2, after 7; tel. 8486.

Kiernan, Jas. G. (R) New York City Univ., 1874; 69 Dearborn, room 17; 10:30 to 1, 3 to 5:30; res. 3226 Forest ave.

King, John Blair Smith (H) Hahnemann, Chic., 1883; 242 Wabash ave., 9 to 1, 2 to 5; tel. 5457; res. 5 Park Row; tel 8455; Prof. of Chemistry and Toxicology, Hahnemann, Chic.

King, Oscar A. (R) Bellevue Med. Col., 1878; 70 Monroe, room 14, 1 to 3; res. Geneva Lake, Wis.; Prof. Diseases of Mind and Nervous System, Col. P. and S., Chic.; Chief Neurological Dept. W. Side Free Disp.; Supt. Oakwood Springs, Lake Geneva, Wis.

King, Wm. (R) Louisville Med. Col.. 1882; 745 S. Halsted, 8 to 10, 1 to 3, after 8; res. 725 S. Halsted; tel. 9021.

Kippax, J. R. (H) Hahnemann, 1869; 3454 Indiana ave., 8 to 9, 1 to 2:30; tel. 8108; Secy. and Prof. Principles and Practice Med. and Med. Jur., Chi. Homeo. Med. Col.

Kirschtein, H. (R) Breslau, 1862; 351 Clark, 8 to 9, 2 to 3; tel. 1401; res. 2921 South Park ave.

Kleene, F. (R) Rush, 1887; 318 Milwaukee ave., 8 to 10, 1 to 3, after 7.

Kleist, H. (R) 103 Clybourn ave., 8 to 10, 5 to 7.

Knoll, Walter F. (H) Chi. Hom. Med., 1879; 726 Washington bv., 8 to 9, 1:30 to 3, 6:30 to 7:30; tel. 7305. Prof. Physiology, Pathology and Minor Surgery, Chi. Homeo. Col.

Knox, J. Suydam (R) P. and S., 1866; 70 Monroe; room 16, 3 to 4; tel. 80; res. 14 Loomis, till 8, 12 to 1, 7 to 8; tel. 7294. Prof. Obstetrics, Rush Med. Col.

Knox, R. C. (R) 1070 W. Harrison; tel. 7201.

Koch, E. P. (R) Missouri Med. Col., 1881; 528 W. Indiana, 8 to 9, 12:30 to 1:30, 3:30 to 4:30, eve.; tel. 7169.

Koehn, J. W. (R) Iowa State Univ., 1885; 501-503 Chicago Opera House bldg., 10 to 12; 860 Milwaukee ave., till 9, 1 to 3, after 7, Sundays till 10; tel. 4573.

Koier, Charles M. (H) Chi. Hom. Med., 1883; N. W. cor. Division and Milwaukee ave., 8 to 10, 2 to 5, 6 to 8; tel. 4518; res. 575 W. North ave.

Koier, Louis G. (H) Chi. Hom. Med., 1888; N. W. cor. Division and Milwaukee ave., 8 to 9, 4 to 5, 8 to 9; tel. 4518; Dispensary hours, Thurs., 1 to 2; res. 575 W. North ave.

Koller, Charles (E) Bennett, 1881; 240 North ave., till 9, 2 to 4, 7 to 9; at 316 Sedgwick; tel. 3247 and 3366.

Kopp, E. (R) Rush, 1884; 90 E. Chi. ave., 8 to 9, 12 to 2, 7 to 8, night; tel. 3511 and 3152.

Korssell, C. F. P. (R) Rush, 1886; 226 Wells, 8 to 9:30, 12 to 2, 6 to 8. Surg. St. Joseph Hosp.; Phys. and Surg. Swedish Home of Mercy.

Kossakowski, M. P. (R) Univ. Mich., 1884; 50 Ingraham, till 9, 1 to 3, 7 to 8; res. 256 W. North ave. Pres. and Prof. Obstetrics Esculapian, Col. of Midwifery.

Kreve, Paul (E) Bennett, 454 North ave., 8 to 10, 2 to 4, 6 to 8; tel. 3206.

Krost, Joseph (R) Rush, 1881; 60 N. Clark; rooms 7 and 9, 8 to 10, 1 to 4, night; tel. 3158 and 3182.

Krueger, J. H. (H) Hahnemann, 1887; 907 35th, 8 to 10, 12 to 1:30, 6 to 7; 360 Blue Island ave.; 2 to 4:30; tel. 9589.

Krusemarck, Chas. (R) Rush, 1879; 125 22d, 9 to 11, 3 to 5, 7:30 to 8:30; tel. 8184; res. 2414 Wabash ave., before 8, 1 p. m, after 8:30; tel. 8031.

Krusemarck, Wm. (R) 125 Clark, 10 to 2; tel. 1339; res. Maplewood.

Kuh, Edwin J. (R) Univ. Heidelberg, 1882; 18 Central Music Hall, 1 to 4:30; tel. 5642; res. 3125 Michigan ave., 7:30 to 8:30, 6 to 7; tel. 8086.

L

La Barriere, Paul E. (R) Rush, 1880; 2906 Archer ave., 8 to 9, 2 to 4; tel. 8195.

OAKWOOD ❋ SPRINGS ❋ SANITARIUM,

LAKE GENEVA, WIS.

HOTEL "LAKESIDE."

THE plans of the Sanitarium include three principal buildings—Hotels "Erb," "Maudsley" and "Lakeside." About fifty patients can at present be accommodated in the two finished Hotels; the plans of which are such as to give to each patient a room, or a suite, as may be required. These communicate with the parlors and halls, so as to afford, in all suitable cases, the utmost freedom and enjoyment of house and grounds. Or, when desirable, many of the most pleasant rooms can be entirely isolated from all others. By these plans, therefore, it is possible to permit the greatest freedom and at the same time to protect every patient in the house from annoyance by any other. Cheerfulness, comfort and safety have been the controlling ideas in the architecture. No crowding is at any time permitted, and the number under treatment will be governed, at all times and entirely, by the best interests of the patients themselves.

In summer we live much out of doors; in winter the buildings are thoroughly warmed in every room, parlor and chamber, from basement to attic, with perfect ventilation, and there is scarcely a room in the entire plan into which the sun may not shine, at some hour during the day. The Sanitarium is not only a summer retreat, but a perfect home of winter comfort.

For further information, class of patients admitted, etc., please see pages 16, 121, and others.

Lackersteen, M. H. (R) King's Col., London, 1858; Univ. of St. Andrew's, Scotland; Royal Col. of Surg., England, 1859; Royal Col. of Physicians, London, 1869; 163 State, 1 to 4; tel. 5480; res. 3962 Cottage Grove ave., 9 to 12, 6 to 8; tel. 9850.

Lackner, E. (E) Bennett, 1872; 2203 Archer ave., 1 to 3, 7 to 9.

Lakey, Anton (H) 378 N. Market.

Landis, E. M. (R) Rush, 1875; 175 Howe, 8 to 9, 12 to 2, night.

Landis, W. H. (R) Starling Med. Col., Columbus, O., 1885; 279 Clark, 8 to 11:30, 4:30 to 9; tel. 1077; res. 254 W. Jackson, 12 to 4; tel. 4311.

Landreth, M. H. (H) Hahnemann, 1885; 2333 Wabash ave., 8 to 10, 7 to 8; tel. 8133.

Lane, J. S. (R) Rock River Med. Soc., Wis., 1847; 232 S. Halsted, 12 to 3, 5 to 8.

Lane, Myron E. (R) P. and S., 1887; 70 State, room 203, 2 to 5.

Laning, C. E. (H) Hahnemann, 1878; Cent. Music Hall, 12:30 to 3; tel. 2642; res. 2972 Calumet ave. Prof. Principles and Pr. Med., Descriptive and Practical Anatomy, Hahnemann.

Larson, C. Frithiof (R) Rush, 1888; 2424 Wentworth ave., 8 to 10, 1 to 3, 6 to 8; Sun., 8 to 9:30; tel. 8099.

Latta, U. G. (R) Col. of P. and S., Chi , 1885; 514 Chi. Opera House, 12 to 2: tel. 9881; res. 3670 Indiana ave., till 9, after 6.

Latimer, H. H. (E) Bennett, 1885; 729 W. Indiana; res., Moreland.

Lawson, L. E. (R) Chi. Med., 1885; 804 W. North ave., 8 to 9, 1 to 2, 6 to 7; tel. 4624.

Leake, Fanny (R) Wom. Med., 1877; office and res., 846 Elk Grove ave., till 10, 4 to 6.

Leavett, Sheldon (H) Hahnemann, 1877; 148 37th, 8 to 9:30, 12:30, 2, 6 to 7:30; tel. 9847. Prof. Medical and Surgical Diseases of Women and Obstetrics, Hahnemann; Clinical Midwifery, Hahnemann Hosp.

Le Caron, Henry (R) Detroit Med. Col., 1872; 177 La Salle ave.

Lee, Edward W. (R) Royal Col. Surg., Ireland; 262 S. Halsted, 1 to 3; tel. 4067; res. 101 Laflin; tel. 7058.

Lee, Julius H. (R) P. & S., 1886; 177 E. Chicago ave., 8 to 10, 1 to 3, after 6; tel. 3274.

Leech, M. S. (R) Eclectic Med. Inst., Cincinnati, 1871; Rush, 1882; S. E. cor. Wabash ave. and Harrison, 8 to 11, 1 to 3; tel. 1448; res. room 9, 367 Wabash av.

Leeming, John (R) Royal Col. P. & S., London, 1886; Col. P. & S., Ontario, Canada, 1886; 4134 S. Halsted, 4:30 to 6; tel. 9646; res. 3400 Indiana ave., 8 to 10, 12 to 2, 6 to 8; tel. 8119.

LeFevre, Wells (H) Chi. Hom. Med., 1887; 56 Cent. Music Hall, 2 to 4; tel. 5241; res. 191 Warren ave., till 9, after 7; tel. 7056.

Leigh, Clarence W. (R) Rush, 1883; 269 Chestnut, till 9, to 2, evenings; tel. 3278. House Phys. and Surg., Chi. Policlinic.

Leininger, Geo. (R) Wooster Univ., Cleveland, O., 1881; cor. Milwaukee ave. and Lincoln, 10 to 12, 7 to 9; res. 764 N. Wood; tel. 4705.

Lemker, H. L. (E) Bennett, 1879; 397 N. Wells, 8 to 10, 1 to 3, 6 to 8.

Lennard, A. L. (H) 3872 Cottage Grove ave., 9 to 11, 2 to 5; res. 3846 Vincennes ave; tel. 9843.

Leonard, Raymond L. (R) Rush, 1869; 94 Wells, 9 to 10, 3 to 4; tel. 3083.

Le Roy, E. W. (H) Hahnemann, Phila., 1883; 130 Dearborn; res. 246 Ogden ave.

Le Roy, W. Gary (E) Bennett, 1881; 130 Dearborn, 9 to 11, 1 to 3; tel. 1339; res. Lombard.

Lewis, Charles J. (R.) Rush, 1875; 733 Carroll ave.; till 9, 12:30 to 1 and 6 to 8.

Lewis, W. F. (R) Rush, 1875; 133 Clark, 4 to 5; tel. 1339; res. 619 W. Monroe, 12 to 2; tel. 7034.

Lewis, W. H. (R) Univ. Mich., 1878; 118 33d, 11 to 12, 5 to 6; tel. 9810; res. 217 53d, till 9, after 6; tel. 9844.

Liebig, Ed. (H) Hom. Col., Dresden, Saxony, 1864; 287 W. 12th, 8:30 to 9:30, 3:30 to 4:30; tel. 4378; res. 962 W. Taylor, till 8, 1 to 3, 6 to 8; tel. 7135.

Lilly, I. N. (R) Louisville, 1864; 422 W. 12th, 8 to 9, 1 to 3, evening.

Lilly, T. A. (R) K'y School Med., 1864; 422 W. 12th, 11 to 12, 3 to 4, 7 to 8.

Lodor, Charles H. (R) Univ. of Penn., 1882; 3136 Indiana ave., before 9, 12 to 1, 5 to 7; tel. 8490.

Loew, Alex. (R) Univ. of Vienna, 1868; 3237 Michigan ave., 1 to 3; tel. 8573.

Lonergan, W. D. (R) Missouri Med. Col., 1879; 199 S. Clark, 8 to 9, 11 to 2, 5 to 7; tel. 1339.

Look, Halleck Hart (R) Col. P. & S., N. Y., 1887; 70 State, 2 to 4; 360 E. Division, 11 to 12; tel. 3284; res. 435 La Salle ave., 9 to 11. Assistant on Nose and Throat, Chi. Polyclinic.

Loomis, Beach (R) Univ. of Vermont, 1870; 113 Madison, 1 to 4; res. 145 S. Robey; tel. 7112.

Lorenz, M. (R) Budweiss, 1872; 719 Allport ave., 8 to 9, 1 to 3, after 6; tel. 9104.

Loring, J. Brown (R) McGill, 1883; Royal Col. S., England, 1884; 674 W. Lake, 10 to 11, 7 to 8; Sun., 1 to 2; tel. 7104.

Lucas, A. C. (R) Univ. of N. Y., 1876; cor. Clark and Van Buren, 8 a. m. to 8 p. m.

Ludlam, E. M. P. (H) Hahnemann, 1861; 70 State, 2 to 4; res. 397 W. Monroe, 8 to 9, 6 to 7; tel. 4195.

Ludlam, Reuben (H) Homeop. Dept. Univ. Penn., 1851; 1823 Michigan ave., 8 to 10, 2 to 4; tel. 8369. Prof. of Med. and Surg., Diseases Women, and Obstetrics, Hahnemann Col.; Diseases Women, Hahnemann Hosp.

Ludlam, Ruben, Jr. (H) Hahnemann, 1886; 1823 Michigan ave., 10 to 12; 4 to 6; tel. 8369.

Luken, M. H. (R) Rush, 1873; 587 W. North ave., till 9, 12 to 2, after 6; tel. 4747.

Lull, Richard (R) Rush, 1883; 234 S. Halsted, till 10, 1 to 2, after 6:30; tel. 4067.

Lundgren, Leonard (H) Chi. Hom. Med., 1881; 29-30 Cent. Music Hall, 10 to 12, 3 to 6.

Lundgren, Sven A. (H) Chi. Hom. Med., 1883; 29 to 30 Cent. Music Hall, 10 to 12, 3 to 6.

Lydston, G. Frank (R) Bellevue Hospital, 1879; 810-814 Chicago Opera House, 10 to 12, 2 to 6; res. 570 Fullerton ave.; tel. 3317. Lecturer Genito-Urinary and Venereal Diseases, Col. P & S.; Surgeon to W. Side Dispensary.

Lyman, Henry M. (R) P. & S., N. Y., 1861; 69 Dearborn st., 1:30 to 4, except Thurs.; res. 533 W. Adams, till 9, after 6; tel. 7130. Prof. Physiology and Diseases of the Nervous System, Rush Med. Col.; Att'g Phys. Presbyterian Hosp. and Central Free Disp.; Prof. Theory and Practice Med., Woman's Med. Col.

Lynch, J. P. (E) Bennett, 1883; 3109 Wentworth ave.

Lyon, H. N. (H) Hahnemann, 1888; 455 Washington bv.; tel. 7298.

Lyon, J. Harvey (R) Univ. Mich., 1878; 221 31st, 9 to 10, 1 to 2:30, 6 to 8; res. 3126 Wabash ave., till 9, 5 to 6.

M

McArthur, Lewis L. (R) Rush, 1880; 600 Bay State bldg., 70 State, 11 to 1; tel. 5558; res. 3305 Forest ave., till 9, after 6; tel. 8227. Prof. Chemistry, Chi. Col. Den. Surg.; Attg. Surg. St. Luke's, Michael Reese, and Chi. Orphan Asylum.

McAuliff, E. L. (R) Rush, 1882; 2202 Michigan ave., till 9, 10 to 12, 3 to 5, 7 to 9; tel. 8184.

McCallister, C. H. (R) Jefferson, 1873; 424 State, 9 to 11, 3 to 5, 7 to 9; tel. 1922; at 103 Adams, 12 to 2.

McCarthy, William (R) McGill, 1867; 171 Blue Island ave., 9 to 10, 2 to 4, 7 to 8; res. 397 W. Taylor.

McCausland, J. W. (R) Rush, 1879; VanBuren and Loomis sts., 10 to 12; 4 to 6. tel. 7213.

McChesney, A. C. (R) Long Island, 1867; 73 Clark; res. 162 Ashland ave.

McClure, V. C. (R) Geneva, 1846; 259 Warren ave., till 9, 12:30 to 2, 6 to 7.
McCormick, F. (H) Chi. Hom. Med., 1884; 516 S. Wood, 8 to 10, 2 to 4, 6 to 7; tel. 7229.
McCoy, J. C. (R) Univ. N. Y., 1887; 224 State; res. 348 Michigan ave.
McCreight, S. Luther (R) Rush, 1885; 559 Ashland ave., till 9, 12 to 2, 6 to 8; tel. 7011.
McCullough, J. R. (R) Toronto, 1858; P. & S., 1883; 9 S. Halsted, 11 to 12, 1 to 3, 7 to 9.
McDonald. Edw. V. (R) Harvard, 1885; 3102 State, 12 to 3, nights; tel. 8166. Mem. Med. Staff Cook Co. Hosp.
McDonald, Peter (R) Rush, 1864; 351 Clark, 3 to 5; res. 2829, Indiana ave.
McDonnell, John A. McGill, Canada, 1870; Buffalo Univ., 1875; Madison and Loomis, till 10, 2 to 4, 6 to 8; tel. 7022; res. 471 Washington bv., 12 to 2.
McFatrich, J. B. (E) Bennett, 1884; Hahnemann, 1885; 126 State, 10 to 5; tel. 9922; res. 3408 Prairie ave.; Prof. Didactic and Clinical Ophthalmology and Otology, Bennett Col.; Ophthalmologist Bennett Hosp.; Prof. Eye and Ear, Bennett Free Disp.
McGaughey, J. A. (R) Chi. Med., 1883; Belvidere bldg., cor. Cottage Grove and 31st, 11 to 12, 3 to 5, 7:30 to 8:30; tel. 8266.
McGinley, J. B. (R) Rush, 1884; 119 E. Madison, 9 to 12:30, 2 to 4:30; res. 778 W. Jackson, 7 to 8, night.
McGorran, M. S. (R) Rush, 1888; Archer ave. and S. Clark, till 11, 2 to 4, after 7; tel. 8081.
McGorray, Chas. H. (R) Michigan Univ., 1882; 2565 Archer ave.; till 9, 2 to 4, nights; tel. 8333.
McGrath, J. J. (R) Rush, 1888; 376 S. Halsted; tel. 4756.
McGrath, M. H. (R) Rush, 1880; 1001 W. Lake, 10 to 12, 4 to 6; tel. 7072.
McGregor, Wm. (R) Univ. N. Y., 1861; 78 S. Sangamon.
McIntosh, L. D. 143 Wabash ave., 8 to 5; res. Ravenswood.
McIntyre, Chas. J. (R) Trinity Med. Col., 1884; 864 W. Indiana, 8 to 10, 12 to 2, 6 to 8; tel. 7279.
McLean, M. G. (R) Rush, 1873; 10 S. Sheldon, 8 to 10, 2 to 4 and 6 to 8.
McLennan, A. S. (R) Royal Col. P. & S., Kingston, Canada, 1873; 277 S. Halsted, room 2, 8 to 10, 2 to 3, 7 to 8; tel. 4067.
McLeod, Edward S. (R) Iowa State Univ., 1877; 68 Randolph, 9 to 8; Sun., 10 to 12.
McNamara, J. R. (R) Col. P. & S., Chi., 1887; 3701 S. Halsted, 9 to 11, 1 to 3, 7 to 9; tel. 9634.
McNeal, D. W. 240 Wabash, 11 to 4; tel. 5457; res. 85 Page; tel. 7148.

McWilliams, S. A. (R) Chi. Med., 1866; 58 State, 11 to 12; res. 3456 Michigan ave., 7 to 8 a. m., 6 to 7 p. m., Sun. 2 to 3; tel. 8157. Prof. Clinical Medicine, Diseases of Chest, and Physical Diagnosis, Col. of P. & S.

MacArthur, R. D. (R) McGill, 1867; 117 S. Clark, 11 to 1, 3 to 5; tel. 3149; res. 414 Dearborn ave., 8 to 9; tel. 3149. Prof. Skin and Venereal Diseases, Chi. Polyclinic; Mem. of Staff of Presbyterian Hosp. and Central Free Disp.

MacGillis, W. C. (R) Royal P. & S., Montreal, 1881; 117 Clark, 8 to 11, 1 to 3, after 5; tel. 1206.

Mace, Abbie J. (R) Wom. Med. Col., 1884; 21 Central Music Hall, 12 to 2; res. 1153 Taylor, 10 to 11, 6 to 7.

Madison, F. M. (R) Col. P. & S., Keokuk, Iowa, 1884; 3212 Graves pl.; tel. 3366.

Maher, J. (R) Rush, 1882; 303 Blue Island ave.

Majeski, Wenzel (E) Breslau, Germany, 1866; Bennett, 1879; 748 Noble.

Major, Laban S. (E) Eclectic Med. Inst., Cincinnati, O., 1848; 951 W. Harrison, till 9, after 4.

Manierre, C. E. (R) Chi. Med., 1882; 229 La Salle ave., 1 to 2:30, 5 to 7; tel. 3292; at 156 Oak, 11 to 12; tel. 3036.

Mannheimer, M. (R) Heidelberg, 1863; 78 State, 1 to 3 p. m.; res. 1822 Indiana ave.

Manning, C. D. (R) Rush, 1880; 163 Washington, 9:30 to 4:30; tel. 592; res., 193 Armitage, 8 to 9, after 4:30.

Marble, W. H. (R) Chi. Med., 1887; 235 Illinois, 9 to 10, 2 to 4, after 7.

Marcusson, W. Beringer (R) Rush, 1885; Harrison and Center ave., 12 to 2, 4 to 5; tel. 4094; res. 429 Carroll ave., till 8, after 6; tel. 7009.

Marguerat, E. (R) Univ., N. Y., 1850; 89 Randolph, 11 to 1, 4 to 5; tel. 1339; res. 708 Monroe; tel. 7056.

Marks, A. J. 339 31st, 12 to 2:30 p. m.; tel. 9824; res. 3847 State, 9 to 11, 3 to 5, evening.

Marley, J. W. (R) 3904 State, 9 to 12, 3 to 4, 7 to 9; res. Englewood.

Marr, W. L. (R) Univ. Mich , 1874; 59 N. State, 10 to 12, 3 to 5, 7 to 8.

Marshall, Ira E. (R) Chi. Med., 1882; 628 W. Lake, 9 to 10, 4 to 5; 7 to 8, night; tel. 7016 Assistant Surg., Ill. Charitable Eye and Ear Disp.

Marshall, John S. (R) Syracuse Univ., N. Y., 1876; "Argyle," cor. Mich. and Jackson, 9 to 4; tel. 2570; res. 3343 Prairie ave. Lecturer on Oral Surg., Chi. Med. Col.; Prof. Oral Surg. and Dean, Univ. Den. Col.; Oral Surg. and Dental Surg., Mercy Hosp. and St. Luke's Free Hosp.

Martin, F. H. (R) Chi. Med., 1880; 163 State, 10 to 1; res. 3308 Rhodes ave.

Martin, H. M. (R) 507 W. Van Buren.

Martin, Louisa (R) Wom. Med., 1886; 729 W. Madison, 10 to 12, 4:30 to 6; tel. 7056.

Martin, William (R) Chi. Med., 1867; 324 Blue Island ave., 11 to 1, 5 to 6, 7 to 8; tel. 4602.

Maschek, F. J. (R) Rush, 1882; 134 Canalport ave.; res. 606 Blue Island ave.

Matter, Martin (R) Chi. Med., 1872; 3123 Wabash, 8 to 9, 1 to 3, 7 to 8; tel. 8564.

Matterson, J. (R) Chi. Med., 1878; 70 State, 1 to 3; tel. 2558; res. 3166 Groveland.

Matthei, Ph. H. (R) Rush, 1861; Goettingen, 1874; 246–248 S. Halsted, 2 to 4; at 136 Canalport ave., 11 to 12; res. 51–53 Wisconsin.

Maul, W. C. (R) Univ. Louisville, Ky., 1869; 3976 Drexel boul., 8 to 9, 1 to 2; tel. 9817.

Mauro, A. G. (R) Univ. Naples, Italy, 1861; 82 W. Madison, 8 to 10, 2 to 4; tel. 4642; res. 196 N. Morgan, till 8, after 6.

Maynard, J. G. (R) Univ. Pa., 1856; 3312 Rhodes ave., 9 to 12, 2 to 5.

Mead, Edward (E) 496 Milwaukee ave.

Meek, J. W. (R) Rush, 1881; 182 Park ave., till 9, 12 to 2:30 and evening. Prof. Chemistry, Am. Col. Dent. Surg.

Meeker, Lysander (R) Keokuk, Iowa, 1857; 134 Van Buren, room 10 Exchange bldg., 10 to 4; res. Cheltenham.

Mellinger, J. H. (R) Univ. Penn., 1874; 10 N. Carpenter.

Mellish, E. J. (R) Rush, 1886; Cook Co. Hosp.; tel. 7133; Interne, Cook Co. Hosp.

Melms, R. (R) Chi. Med., 1881; 659 W. 12th, 7 to 9, 2 to 4, 7 to 9; tel. 7011.

Mentzel, Carl Meyers (R) Univ. Berlin, 1853; 23 Eugenie, 8 to 9, 3 to 7; tel. 3208.

Mercer, Frederick W. (R) P. & S., N. Y., 1862; 37 26th; res. 2600 Calumet, 9 to 10, 1 to 2, 7; tel. 8524.

Merckle, H. (R) Wurzburg, 1858; 1528 Michigan ave., 1 to 3; tel. 8059.

Mergler, Marie J. (R) Wom. Hosp. Med., 1879; Central Music Hall, 11 to 12:30; tel. 2558; res. 29 Waverly pl. Secy. and Prof. Gynæcology, Wom. Med. Col.

Merrick, J. C. (R) Rush, 1867; Garfield House, cor. Armitage and Milwaukee aves, 9:30 to 10:30, 1:30 to 3, after 7; res. 1061 N. Western ave., 8 to 9, 12 to 1, 5:30 to 6:30.

Merrill, S. B. L. (R) P. & S., 1886; 98 Harrison, 11 to 1, 4 to 6, 7 to 8; tel. 1642; res. cor. Throop and 12th, 8 to 10, 2 to 4, night.

Merriman, Henry P. (R) Chi. Med., 1865; 2239 Michigan ave., 11 to 1; tel. 8047. Lecturer on Gynæcology, Rush Med. Col.

Messinger, C. D. (E) Bennett, 1883; Central Music Hall, 1 to 5; res. 913 W. Monroe.

Messinger, E. D. (E) Bennett, 1887; Central Music Hall, 3 to 4; tel. 2642; res., 913 W. Monroe, 9 to 11, 5 to 8; tel. 7195.

Meyer, Baltazar (R) Univ. of Norway, 1877; 241 Milwaukee ave., 12:30 to 1:30, 6 to 7; res. 60 Fowler, till 10, 3 to 6; tel. 4203.

Meyer, Friedrich B. (R) Berne, 1840; 56 W. 19th, 8 to 10, 2 to 3, after 6; tel. 9021.

Meyer, Lucy Rider (R) Woman's Hosp. Med. Col., 1887; 114 Dearborn ave., Chi. Training School for Missions.

Michelet, W. E. J. (R) Rush, 1879; 503 W. 12th, till 9:30, 12 to 2:30, after 6.

Miessler, E. G. H. (H) Hahnemann, 1873; 737 S. Halsted, 8 to 10, 1 to 3, 6 to 8; tel. 9021.

Miller, Adam (H) N. Y. Univ., 1847; 172 Ashland ave., 8 to 10, 1 to 3.

Miller, Bruce (R) P. & S., 1886; 3552 S. Halsted, 9 to 11, 3 to 5, 7 to 8; tel. 9621; res. 3616 La Salle; till 8:30, 12 to 2, 5:30 to 6:30, after 8.

Miller, B. C. (R) Rush, 1868; 3033 Groveland ave., till 10, 3 to 6, at night; tel. 8354.

Miller, De Laskie (R) Geneva, N. Y., 1842; 2011 Prairie ave., 12 to 2. Prof. Obstetrics and Diseases of Children, Rush Med. Col.; Attending Phys. for Diseases of Children and Accouchers, Presbyterian Hosp.; Attending Obstetrician St. Luke's Free Hosp.; Consulting Phys. Woman's Hosp.; Phys. Home for Incurables.

Miller, Geo. W. (R) 2719 State.

Miller, O. G. (R) Chi. Med. Col., 1888; 82 Wells, 9 to 11, 2 to 4, after 7; tel. 3426.

Miller, Truman W. (R) Geneva, N. Y., 1863; 211 Opera House bldg., 12 to 3; res. 1071 N. Clark, till 8, 5 to 7; tel. 3213.

Miller, William E. (R) P. & S., 1887; 900 W. 21st, 10 to 11, 3 to 4; tel. 9204; res. 1111 S. California ave., till 9, 12 to 2, after 6; tel. 9058.

Mills, J. Lee (R) Univ. Mich., 1860; 38th and Archer ave., 7 to 9, 12 to 1, 5 to 9; tel. 9505.

Mills, James P. (H) Hahnemann, Philadelphia, 1874; 903 Monroe, 8 to 9, 1:30 to 3:30, 6 to 7; tel. 7050.

Mintie, R. L. (E) Bennett, 1881; 182 W. Madison, room 6.

Mitchell, Clifford (H) Chi. Hom., 1878; 70 State; 4 to 6; 44 16th. Prof. Chemistry and Toxicology, Chi. Hom. Med. Col.

Mitchell, J. S. (H) Bellevue, 1865; 2954 Prairie ave., 8 to 9, 2 to 4 tel. 8702. Pres. and Prof. Institutes and Prac. of Med., Chi. Hom. Med. Col.

Mitter, Robert (R) Rush, 1882; 387 Blue Island ave., 8 to 10, 1 to 3, after 7; tel. 4160.

Mixer, Mary A. (R) Wom. Med., 1884; 1931 Indiana ave., 9 to 12; tel. 8332. Lecturer on Histology, Woman's Med. Col.; Staff of Hosp. for Women and Children.

Montgomery, Frank H. (R) Rush, 1888; 240 Wabash ave., 10 to 2; res. 193 S. Lincoln.

Montgomery, L. B. 118 Park ave., morning and evening.

Montgomery, Liston H. (R) Chi. Med., 1871; 189 Randolph, 2 to 6, Sunday 12 to 2; tel. 1163; res. 118 Park ave., till 8, after 6; tel. 7066. Med. Inspector, N. W. Division.

Montgomery, W. A. D. (R) Victoria Med., 1881; 131 Dearborn ave., till 9, 12:30 to 2, and 6 to 7:30; tel. 3125.

Montgomery, William T. (R) Rush, 1871; 112 S. Clark, 10 to 1; res. 592 W. Van Buren. Prof. Ophthalmology and Otology, Woman's Med. Col.; Attg. Surg., Ill. Char. Eye and Ear Infirmary; Attg. Phys., Central Free Disp.

Moore, Blanche (R) Wom. Hosp. Med., 1886; 8 Central Music Hall, 2 to 4; res. cor. Garfield and Lincoln aves., till 8, 10:30 to 1, 5 to 7; tel. 3634.

Moore, C. F. (R) Rush, 1888; 254 S. Halsted, 9 to 12, 6 to 8; tel. 4067.

Moore, D. G. (R) Rush, 1879; 721 Elston ave., 9 to 10, 1:30 to 3, 5 to 6; res. 643 N. Hoyne, 12 to 1:30, after 6; tel. 4648.

Moore, French (R) Rush, 1880; 1536 Ogden ave., till 9, 12 to 1, after 6.

Moore, Willis F. (R) Rush, 1880; 3835 State, till 9, 2 to 4, 7 to 8; tel. 9824; at 2204 Archer ave., 10 to 11.

Moran, M. Connor (R) Rush, 1882; 220 Milwaukee ave., 8 to 10, 2 to 4, and 7 to 9; tel. 4612.

Morey, A. C. (A) Jefferson, 1857; 182 W. 12th, 4 to 5 p. m.; tel. 4049; res. 534 Claremont ave., till 9, 12 to 3, evening till 7:30; tel. 7289.

Morey, N. Derexa (R) Wom. Med., Chi., 1885; 885 W. Adams, till 9, 10:30 to 12:30, 6 to 7; tel. 7195. Asst. to Chair of Physiology, Wom. Med. Col.

Morgan, William Edward (R) Chi. Med., 1882; 3160 State, 9 to 10, 5 to 6; tel. 8330; res. 34 E. 29th, 12 to 2; tel. 8210. Demonstrator Operative Surgery, Chi. Med. Col.; Surgeon, South Side Disp.

Morgan, W. H. (R) Rush, 1874; cor. Center ave. and 14th, 10:30 to 12; res. 454 W. Jackson, till 9, 12 to 2 p. m., evening; tel. 4089. Clinical Asst. Chair of Diseases Children, Rush Med. Col.

Morin, Denis (H) Hahnemann, 1884; 172 Blue Island ave., 9 to 11:30, 3 to 5 and evening; tel. 4673; res. 109 Blue Island ave.

Morrison, J. P. (R) Univ. Mich., 1868; 3030 Michigan ave.

Morris, Henrietta K. (E) Bennett, 1885; Central Music Hall, 2 to 5; tel. 5642; res. 300 Washington boul.

Mott, J. W. (R) Rush; 12th and Ashland ave., 8 to 9, 1 to 3.

Moyer, Harold (R) Rush, 1879; 434 W. Adams, till 9, 12:30 to 1:30, 6 to 7:30; tel. 4703. County Phys. Cook Co., Asst. Chair Diseases Nervous System, Rush Med. Col.

Mueller, Ida (R) Wittenberg, Germany, 1880; 651 W. Monroe.

Mulfinger, J. L. (R) Rush, 1880; 583 Halsted, 10 to 10:30, 1 to 3, 7 to 8; tel. 9006; res. 942 W. Jackson, 8 to 9, 5 to 6, after 9; tel. 7091.

Mullan, E. A. (R) Jefferson, 1874; 722 W. 21st, till 9, 12 to 2 and evenings; tel. 9151.

Munn, Katherine D. (E) Bennett, 1882; 243 State, 3 to 5; res. 3728 Dearborn.

Munsell, Anson (R) Chi. Med., 1871; 2724 State.

Munzer, Isador (R) P. & S., 1884; 60 Canalport ave., till 10, 12 to 2, after 5.

Murdock, E. P. (R) Rush, 1880; 161 W. Madison, 3 to 4; res. 148 Loomis, till 8, 1 to 2, evening; tel. 7179. Lecturer on Gynæcology Col. P. & S., Chi.; Lecturer on Throat and Voice, Am. Conservatory of Music.

Murphy, John B. (R) Rush, 1879; 262 S. Halsted, 1 to 3; tel. 4067; res. 36 Throop.

N

Neel, W. D. (R) Univ. of Louisville, 1879; Erie and Noble, till 9, 12 to 1:30, 5:30 to 7:30; tel. 4564.

Neill, W. J. (R) Rush, 1880; 126 Milwaukee ave., 2 to 3, 7 to 8, tel. 4288; res. 296 N. Lincoln, till 9, 5 to 6, nights; tel. 7087.

Nelson, Daniel T. (R) Amherst, 1864; Harvard, 1865; 125 State, 10:30 to 12:30; res. 2400 Indiana ave., 8 to 8:30, 7 to 7:30; tel. 8107. Adj. Prof. Gynæcology, Rush Med. Col.; Surg. Woman's Hosp.

Newman, Henry Parker (R) Detroit Med. Col., 1878; 65 Randolph, 12 to 2; tel. 5558; res. 554 W. Monroe, till 8, 6 to 7; tel. 7199. Lecturer on Obstetrics, Prof. Diseases of Children, Col. P. & S.; Gynæcologist, Chi. Policlinic; Gynæcologist, West Side Free Disp.

Newton, George W. (R) Univ. Penn., 1884; 197 E. Madison, 11 to 1; tel. 1324; res. 971 W. Van Buren, till 9, 2 to 3:30, after 6; tel. 7126.

Nichols, Stella B. (R) Wom. Med. Col., 1882; 230 Ogden ave., 10 to 2; tel. 7300.

Nichols, W. T. (E) Bennett, 1883; 522 W. Monroe, 9 to 4.

Nielsen, N. J. (R) Chi. Med., 1878; 39 Ray ave., 10 to 11, 2 to 3, 7 to 9.

Nielsen, Theodore (H) Chi. Homeo. Med. 1886; 322 W. Chicago ave., 8 to 10, 1 to 3, 7 to 8; tel. 4646; res. 322 W. Chicago ave.

Niles, J. W. (R) Univ. City of N. Y., 1876; 396 La Salle ave., 12 to 2, 7 to 8; tel. 3611.

Nitz, C. F. (R) Rush, 1880; 800 N. Halsted, 7 to 9, 1 to 2, 6 to 8; at 136 Fullerton ave., 4 to 5.

Nolan, D. W. (R) Chi. Med., 1868; 476 Milwaukee ave., 10 to 12, 4 to 6, 7 to 9; tel. 4029.

Norcom, F. B. (R) Univ. N. Y., 1854; 3640 Cottage Grove, 9 to 10, 7 to 8; res. 3638 Vincennes ave.

Novak, Frank J. (R) Rush, 1885; cor. 12th and Halsted, 11 to 12; tel. 4378; res. 731 W. 18th, 8 to 10, 12 to 2, after 7; tel. 9078.

O

O'Connell, Patrick (R) Queen's, Ireland, 1872; 439 W. Taylor, till 9, 2 to 4, evenings; tel. 4071.

O'Connell, Wm. (R) Queen's Col., Ireland, 1849; 126 Milwaukee ave., 10 to 12, 2 to 4, 6 to 9; res. 150 N. Peoria; tel. 4288.

O'Malley, T. F. (R) Rush, 1886; 171 Blue Island ave., 9 to 10, 2 to 3, 7 to 9; tel. 4561.

O'Ryan, C. D. B. (R) 248 N. Wells.

O'Shea, David (R) Rush, 1883; 709 W. 21st, till 9, 12 to 2, at night; tel. 9151.

Ochsner, Albert J. (R) Rush, 1886; 300 S. Wood. Demonstrator Physiology and Pathology, Rush Med. Col.

Ogden, E. J. (R) Univ. of City of N. Y., 1855; 70 Monroe, 11 to 1; res. 1636 Michigan ave., 8 to 9, 2 to 3, 6 to 7 tel. 8511.

Ogden, E. Russell (R) Chi. Med. 1884; 70 Monroe, 9 to 11; res. 1800 Indiana ave., 8 to 9, 6 to 7; tel. 8511.

Ogden, M. D. (H) Hahnemann, 1864; 163 State, 1:30 to 3:30; res. room 255 Palmer House; tel. 5480.

Ohlendorf, W. C. (R) Chi. Med., 1882; 645 Blue Island ave., 7 to 9, 12 to 2, 7 to 9; tel. 9086.

Olin, A. C. (E) Bennett, 1876; 112 E. Van Buren, 8 a. m. to 8 p. m.; tel. 1155.

Olin, Henry (E) Eclectic Med. Col., Penn., 1860; 163 State, 10 to 4; tel. 5480; res. 3619 Indiana ave.

Oliver, Thomas Telfer (H) Rolph Med. School, 1858; 2306 Indiana ave., 1 to 2, 7 to 9; tel. 8060.

Orgleit, Moriah T. (R) Boston Univ., 1881; Univ. Mich., 1884; 562 Noble.

Orr, Julia M. (H) Hahnemann, 1887; 3411 Cottage Grove ave.

Oswald, Julius W. (R) Rush, 1887; tel. 3467; Resident Phys., Alexian Bros. Hosp.

Otto, Julius (R) Chi. Med., 1876; Rush, 1877; Univ. of Goettingen, 1879; 125 State, 2 to 4; tel. 5442; res. 468 W. Chicago ave., till 8:30, 12 to 1, 6 to 7; tel. 4304.

Owens, John E. (R) Jefferson, 1862; 1806 Michigan ave., 11 to 1, 7 to 8; tel. 8216. Prof. of Surgical Anatomy, Operative Surg. and Clin. Surg., Chi. Med. Col.; Attg. Surg. St. Luke's Free Hosp.

P

Page, J. M. (E) 49 N. May, 9 to 12, 6 to 9; tel. 4193.

Pague, Chas. (R) McGill, 1867; 235 State, 10:30 to 4; res. 18 Lincoln ave.

Painter, Dayton (R) Rush, 1876; 1412 Wabash ave., 9 to 10, 1 to 3, 7 to 9.

Palmer, Thomas D. (R) Chi. Med., 1867; Jefferson, 1882; 259 Warren ave., 8 to 10, 1 to 3, 5 to 6; tel. 7049; res. 243 Park ave.; tel. 7094.

Paoli, Gerhard C. (R) Rush, 1866; 26 N. Clark, 1 to 3; res. 1834 Frederick, 8 to 9, 5 to 6; tel. 3352. Consulting Phys. Chi. Hosp. for Women and Children; Prof. Emeritus Materia Medica and Therapeutics, Woman's Med. Col.; same in N. W. Col. of Den. Surgery.

Park, Andrew James (R) Harvard, 1852; Univ. of Victoria, Ont., 1858; Col. P. & S., N. Y.; 3200 Vernon ave., 8 to 10; 110 Dearborn, 2 to 4.

Park, Augustus V. (R) Rush, 1883; 420 26th, 9 to 11 a. m., 1 to 3, 6 to 8 p. m.; tel. 8372.

Park, Lottie J. (R) Wom. Med., 1881; Central Music Hall, 9 to 10; res. 456 N. Clark, 11 to 1, 4 to 6.

Parker, Anna M. Chi. Hom., 1878; 28 Grant pl., 9 to 12, 6 to 7.

Parkes, Charles T. (R) Rush, 1868; 125 State, 1:30 to 2:30; tel. 5442; res. 51 Lincoln ave., 11 to 12, 5 to 6; tel. 3246. Prof. of Surgery, Rush Med. Col.; Lecturer Ill. Col. of Pharmacy; Attg. Surg. Presbyterian Hosp.; Surg. in Chg. St. Joseph's Hosp.; Phys. for Home for Incurables.

Parsons, Geo. F. (R) P. & S., 1876; 3904 Cottage Grove, 8 to 10, 3 to 6, 7 to 8:30; res. 3933 Drexel bv.; tel. 9817.

Patera, F. J. (R) Rush, 1883; 709 Milwaukee ave., 3 to 4:30; tel. 4333; 573 Blue Island ave., 10 to 11:30; res. 144 W. Taylor.

Patrick, Hugh T. (R) Bellevue, 1884; 307 Division, till 9, 4 to 5, 7 to 8, night; tel. 3284; res. 426 N. State, 5 to 7; tel. 3552. Attg. Phys. to Disp. of Chi. Policlinic.

Patton, Joseph M. (R) Univ. N. Y., 1882; 237 S. Hoyne ave.; tel. 7188.

Paul, Ph. D. (H) Hahnemann Med. Col. Chi. 1883; 343 N. Clark, 12 to 2, 6 to 7.

Pearman, Jas. O. (R) Rush, 1885; 161 W. Madison.

Pearson, N. P. (R) Univ. Copenhagen, 1853; 89 Randolph, 1:30 to 3; res. 655 LaSalle ave., 8 to 9, 7 to 8.

Peiro, F. L. (H) Vermont Univ., 1866; 78 State, 9 to 5; res. 171 State.

Perkins, Charles F. (R) Rush, 1886; 82 W. Madison, 9 to 11, 2 to 5, 7 to 8; tel. 4642.

Pettengill, J. B. (R) Univ. Penn., 1870; 2503 Indiana ave., 8 to 9, 6 to 8; tel. 8060.

Pettyjohn, Elmore S. (R) Rush, 1882; 828 W. Madison, till 9, 12 to 3, 6 to 7; tel. 7128. Prof. Clinic Department, Rush Med. Col.

Phelan, J. B. (R) McGill, 1874; P. & S., of Ontario, 1876; 477 W. Indiana, 9 to 11, 2 to 4, Sun. 1 to 3; tel. 7157.

Phelon, W. P. (R) Univ. Iowa, 1865; 629 Fulton: till 11:30.

Phillips, H. A. (E) Bennett, 50 W. Madison, 9 to 11, 1 to 3, 8 to 9, Sun. 9 to 11 and 1 to 2:30; tel. 4475; res. 844 W. Adams; tel. 7176.

Phillips, M. (R) Victoria, 1862; 159 Clark.

Pickard, J. C. (R) Rush, 1887; 212 Milwaukee ave., till 9, 12 to 2, 5 to 6, at night; tel. 4612.

Pierce, Norval H. (R) P. & S., 1884; cor. Randolph and State, 9 to 10, 3 to 6; tel. 5558; res. 49 Rush, 12 to 2, evenings; tel. 3316.

Pillsbury, J. M. (R) Jefferson, 1865; 3239 Calumet ave.

Pingree, M. Gaylord (E) Bennett, 1880; 103 State, 11 to 3; res. Ravenswood, 6 to 8.

Piper, R. J. (R) Univ. Mich., 1875; 312 W. Indiana, 10 to 11, 2 to 3, 7 to 8, Sun. 10 to 12; tel. 4235.

Pischczak, John Lemberg, Aus., 1862; 2904 Vernon ave., till 9, 1 to 4, after 7.

Plaum, Charles (R) License Baden, 1875; 2901 Wentworth ave., 7 to 10, 1 to 3, after 7; tel. 8296.

Plecker, J. H. (R) Rush, 1877; 183 W. Madison, 10 to 11, 2 to 6, 7 to 8; tel. 4148; res. 320 W. Adams, till 10, after 8; tel. 4352.

Plumb, Henry (R) Yale, 1861; 240 Wabash ave., 10 to 4; tel. 5457; res. 90 Loomis; tel. 7213.

Plummer, Charles G. (R) Chi. Med., 1886; Continental Hotel; Wabash ave. and Madison. till 10 at night.

Poppe, Otto (H) Hahnemann, 1870; 4441 Wentworth ave., 11 to 12; res. 2727 Portland ave., till 9, 1 to 2, after 6; tel. 8032.

Porter, Fred D. (R) Detroit Med. Col., 1877; 1594 N. Halsted, 7 to 9, 12 to 1, 5 to 7; tel. 3919. Surg. Maurice Porter Mem. Hosp.

Porter, M. N. (R) Bowdoin, 1880; cor. State and 39th; tel. 9624.

Powell, Edwin (R) Rush, 1858; 41 Clark, 9 to 12.

Pratt, E. H. (H) Hahnemann, 1873; Cent. Music Hall, 2 to 4; tel. 5241; res. 519 LaSalle ave. Prof. of Principles and Practice of Surg., Chi. Homeo. Med. Col.; Surgical Dept., Cent. Homeo. Hosp. and Free Disp.

Pratt, Leonard (H) Homeo. Med. Col., Penn., 1862; Cent. Music Hall, Wednesdays and Saturdays, 10 to 2; tel. 5241; res. Wheaton, Ill.

Prestley, James P. (R) Rush, 1886; 125 State, 1 to 3; tel. 5442; res. 310 S. Hoyne, till 9, 6 to 7; tel. 7188.

Price, Oscar J. (R) Univ. of Mich., 1866; cor. Halsted and Van Buren, 10 to 11, 2 to 4, 7:30 to 8:30, Sun. 10 to 11; tel. 4640; res. 625 W. Jackson, 8 to 9, 12 to 1, 6 to 7; tel. 7219. Mem. Surg. Staff Cook Co. Hosp.

Priestman, J. (R) Downing, Eng., 1855; 3640 Cottage Grove, 10 to 12, 2 to 4; tel. 9874; res. 3401 Vernon.

Prince, Isaac (H) Hannemann; Chi. Foundling's Home, 9 to 5; res. 114 S. Wood. Res. Phys. Chi. Found ling's Home.

Prince, Lawrence H. (R) Rush, 1885; 51 Lincoln ave., 12:30 to 2; tel. 3246; res. 78 Lincoln ave., 11:30 to 12:30, after 8; tel. 3669.

Printy, J. A. (H) Homeo. Med. Dept. Univ. of Iowa, 1882; 598 Lincoln ave., till 9, 12:30 to 2, after 6; tel. 12,035.

Profeck, J. W. (E) Am. Col., St. Louis, 1882; 594 Canal, 8 to 10, 1:30 to 3:30 and nights; tel. 9076.

Provost, William Y. (R) Long Island Hosp. Col., N. Y., 1877; 430 W. Randolph, 12 to 4, 8:30 to 9:30; tel. 7125; res. 274 Loomis, 8 to 10; tel. 7110.

Purdy, Chas. W. (R) Royal Col. P. & S., Ontario, Queen's Univ., Kingston, 1869; 163 State, 10 to 1, 2 to 3; res. Grand Pacific Hotel.

Pusheck, Charles (H) Hahnemann, 1879; 330 LaSalle, 11 to 5.

Pynchon, Edwin (R) Med. Col. of Ohio, 1876; 703 Opera House blk., 11 to 1; tel. 1206; res. 561 W. Madison, 8 to 9, 4 to 6; tel. 7172.

Q

Quales, Niles T. (R) Rush, 1866; 241 Milwaukee ave., 9 to 10, 2 to 3, 7 to 8; res. 52 Fowler, till 9, 12 to 1.

Quine, William E. (R) Chi. Med., 1869; 58 State, 12:30 to 2; tel. 5558; res. 32d st. and Indiana ave., 8 to 9; tel. 2558. Prof. Practice of Med. and Clinical Med. Col P. & S, Chi.; Prof. Materia Medica, Ill. Col. Pharmacy.

R

Rahlfs, Th. (R) Rush, 1888; 527 S. Halsted; tel. 9139.
Rakenius, H. (R) Rush, 1878; 405 Larrabee, 1 to 5; tel. 3366; res. 288 North ave. till 10, after 6.
Randell, George H. (R) Univ. Mich., 1878; 4024 Drexel boul., 8 to 9, 4 to 5; tel. 9809.
Randolph, Robert (R) Bellevue, 1879; 2139 Wabash ave., 8 to 10, 3 to 5, 7 to 8.
Rankin, A. C. (R) Starling Med. Col., Columbus, O., 1852 ; 202 State, 2 to 5; res. 9 Market, Pullman.
Rauch, John H. (R) Univ. Pa., 1849 ; Sec'y State Board Health, Springfield ; Grand Pacific, Chicago, Sat. and Sun.
Re, Nicholas (R) Naples, 1880 ; 359 Dearborn, 7 to 9; 12 to 3, 6 to 10; tel. 1401.
Rea, R. L. (R) Med. Col. of Cincinnati, 1855; 272 E. Huron, 12 to 2; Sun., 1 to 2.
Reading, Edgar (E) Indiana Med. Col., 1849; Univ. N. Y., 1857 ; 3750 Langley ave., 8 to 10, 12 to 2, 7 to 9.
Reading, Edgar M. (E) Bennett, 1877 ; 103 State, 1 to 4 ; tel. 3453; res. 3748 Langley ave., 7 to 9, 7 to 9; tel. 9874. Prof. Diseases of Respiratory and Circulatory Organs and Nervous System, Bennett Col.
Reasner, Marie E. (E) Bennett, 1878: 23 Central Music Hall, 11 to 2; tel. 5642; res. 5729 LaSalle, till 9; 5 to 6, Englewood; tel. 20.
Reed, Chas. Bert. (R) Rush, 1887 ; 240 Wabash ave., 9 to 5; tel. 5457; res. 41 Seeley ave., nights; tel. 7049.
Reed, W. E. (H) N. Y. Hom. Med. Col., 1884 ; 16-17 Central Music Hall, 1 to 4; tel. 5642 ; res. 427 LaSalle ave., 8 to 9, 7 to 8; tel. 3312. Diseases of Children, Hahnemann Col.
Reed, T. J. (R) Virginia, 1861; 807 Chi. Opera House, 10:30 to 4; tel. 1294; res. Ravenswood.
Reeder, Wm. DeH. (R) Nashville, 1865; 114 Honore.
Reilly, Frank W. (R) Chi. Med., 1861; 682 W. Adams. Demonstrator of Anatomy, Chi. Med. Col.
Reilly, Joseph (R) St. Louis Med. Col., 1872; 3835 S. Halsted, 9 to 10. 2 to 3, 8 to 9; tel. 9640.
Reinhold, W. (R) Rush, 1868; 146 N. Clark, 10 to 12.
Reininger, E. E. (H) Chi. Hom., Med., 1888; 1093 W. Taylor 8 to 10 6 to 8 tel. 7289, at 903 S. Ashland ave. 1 to 2.
Remmen, N. E. (R) P. and S., Chi., 1887; 114 N. Center ave., 8 to 10, 6 to 8.
Reynolds, Arthur R. (R) Bellevue, N. Y., 1876 ; 229 E. Division, 8 to 10, 2:30 to 6:30; tel. 3049.
Reynolds, Mrs. Belle L (H) Hahnemann, 1880; 1823 Michigan ave.
Reynolds, Ben Philips (R) Bartholomew, London, 1857 ; 128-130 Clark, 10 to 12, 3 to 5; tel. 1339; res. Narberth Castle, Lombard.

Reynolds, George Warren (R) Rush, 1873; 232 Milwaukee ave., 10 to 12, 3 to 4, 7 to 8 ; tel. 4011 ; res. 335 Washington boul., till 9, 1 to 2, nights ; tel. 4656.

Reynolds, Henry J. (R) Bellevue, N. Y., 1875; 163 State, 9:30 to 11, 5 to 6, 7 to 8 ; tel. 5480 ; res. Palmer House. Prof. Dermatol., Col. P. and S.; Prof. Skin and Genito Urinary Diseases, Chicago Policlinic; Chief Dermatologist and Genito Urinary Surg., W. Side Disp., Chicago.

Reynolds, J. E. (R) Central Col. P. & S., 1885; 336 W. Twelfth, 10 to 12, 7 to 8; res. 429 S. Oakley, 5 to 6, night; tel. 7024.

Revkowski, Casimir, (R) Charkow, Russia, 1873; 615 Noble, till 9, 12 to 3, evenings; tel. 4400.

Rhodes, John Edwin (R) Rush, 1885; 70 State, 10 to 2; tel. 5558; 132 S. Halsted, 8 to 9, 4 to 5, 7 to 9 ; res. 488 W. Congress, till 7:30, 5 to 6:30 ; tel. 7001. Att'g Phys. to Central Free Disp., Throat and Chest Dept.

Rice, N. B. (R) Albany. 1854; 222 Marshfield ave.

Richards, Geo. E. (H) Hahnemann, 1878; Central Music Hall, 3 to 5; tel. 2642; res. Clarendon Hotel, till 9, 1 to 2, 6 to 7:30; tel. 3166.

Richardson, J. R. (R) McGill, 1865; 3015 Lock, till 9, 1 to 2, 7 to 8; tel. 8271.

Ring, John (R) Rush, 1888; 229 W. Randolph, till 9, 12 to 1:30, 5:30 to 7:30; tel. 4270.

Rittenhouse, William (R) P. & S., 1886 ; 987 Ogdon ave., 11 to 12 ; tel. 7309; res. 479 S. Leavitt, 9 to 10, 1 to 2, 6:30 to 7:30; tel. 7184.

Rivenburg, E. L. (R) S. E. cor. Washington and 5th ave., room 17, 9 to 5 ; Tues. and Thurs., 7 to 9 ; res. 64 Grant place.

Robertson, Jessie E. (H) Hahnemann, 1886 ; 2612 S. Park ave., 9 to 12.

Robison, J. A. (R) Rush, 1880; 70 State, 2 to 3:30; tel. 5558 ; res. 428 Washington boul., 12 to 1:30 and evenings; tel. 7292. Lecturer on Materia Medica, Rush Med. Col.; Sec. and Attndg. Phys. for Throat Diseases, Presbyterian Hospital ; Phys. Cook Co. Hospital ; Attending Phys. Central Free Dispensary ; Prof. Materia Medica and Therapeutics, Woman's Med. Col.

Robinson, Wm. (R) Buffalo, 1862; 114 LaSalle, 10 to 12, 1 to 3; res. 84 Johnson ave., 7 to 8, 5 to 8.

Rockwell, H. O. (R) Chi. Med., 1881; 144 Oakwood boul., 10 to 12, 3 to 5, 8 to 9; tel. 9966; res. 3805 Johnson place, 8 to 9, 1 to 2.

Roesch, Fred (H) Heidelberg, Germany, 1850; 113 Adams, cor. Clark, 12:30 to 3; res. 545 LaSalle ave., till 9, 6 to 8; tel. 3027.

Roesenburg, A. G., (R) Rush, 1883; 89 Randolph, 9 to 11, 1 to 3, 6 to 8; res. 458 Cleveland ave.

Rogers, B. W. (R) P. & S., 1885; 878 W. Adams; tel. 7243.

Rogers, Fred D (R) Chi. Med., 1880; 2412 Cottage Grove ave., 7 to 10, evenings; tel. 8174.

Rogers, L. D. (H) Hahnemann, 1884; 441 Dearborn ave., till 9, 1 to 3, 6 to 8; tel. 3169.

Rogers, S. Ida Wright (H) Hahnemann, 1884; 441 Dearborn ave., till 9, 1 to 3, 6 to 8; tel. 3169.

Roler, E. O. F. (R) Rush, 1859; 125 22d, 1 to 3; res. Southern Hotel. Emeritus Prof. Obst. Chi. Med. Col.; Consulting Phys. Mercy Hospital.

Ronga, G. (R) Naples, 1873; 418 Clark.

Roop, J. E. Phys. Med. Inst., Cincinnati, 1866; 1190 W. Harrison, till 9, 5 to 7:30.

Root, Mrs. E. H. (R) Wom. Med., 1882; 26 Central Music Hall, 1 to 2, except Thurs.; tel. 2558; res. 596 W. Adams; tel. 7037. Prof. Hygiene, Chi. Training School; Attendant Obstetrical Dept. Chi. Hospital for Women and Children; Prof. of Hygiene and Med. Jurisprudence, Woman's Med. Col.

Rosenberg, Adolph (R) Berlin, — ; 89 Randolph, 9 to 11, 1 to 3, 6 to 8: res., 458 Cleveland ave.

Rosenberg, N. (R) Norway, 1852; 41 S. Clark, 9 to 4; res. 34 Evergreen ave., after 6.

Rosenthal, Adolph (R) Vienna, 1869; 707 Milwaukee ave., 7 to 9, 1 to 3; res. 850 W. Monroe.

Rosenthal, D. (R) Univ. N. Y. City, 1879; 802 Larrabee, 7 to 9, 1 to 3, 6 to 8; tel. 3408.

Rosenthal, Jacob (R) Jefferson, 1888; 2871 Archer ave., 9 to 10, 2 to 3, night; tel. 8195; res. 3014 Deering, till 9, 1 to 2, 6 to 7.

Ross, Joseph P. (R) Ohio Med. Col., 1851; 70 State, 9 to 2; res. 428 Washington boul.; tel. 7292. Prof. Clinical Med. and Diseases of Chest, Rush Med. Col.; Att'g Physician Presbyterian Hosp., Cook Co. Hosp., and Central Free Disp.

Rowan, Peter J. (R) Toronto Univ., 1870; 160 W. Harrison, 11 to 11:30, 5 to 5:30; res. 138 S. Desplaines, 1 to 3, 7 to 8; tel. 4055.

Ruggles, Georgia Sackett (R) Wom. Med., 1883; 2211 Michigan ave., till 9, 1 to 2:30; tel. 8172. Att'g Phys. Erring Woman's Refuge.

Runnels, John F. (R) Louisville Med., 1879; 178 Seminary ave., till 9, 12 to 2, 6 to 8; tel. 3299.

Ruschhaupt, H. F. (R) Berlin, 1852; 135 Clybourn ave.; tel. 3152.

Rush, Edwin F. (E) Bennett, 1870; 103 State.

Rutherford, C. (R) McGill, 1882; 102 Fullerton ave., till 9:30; 1 to 3, after 7; tel. 3924. Prof. of Anatomy, Phys. and Surg. Hosp.; Att'g Phys. Chi. Policlinic.

Rutter, C. O. (R) Chi Med, 1868; 432 Dearborn ave., 11 to 12; tel. 3410.

Ryan, E. P. (R) Rush, 1886; 200 Chicago ave., 10 to 12, 2 to 6.

Ryan, J. E. (E) Bennett, 1882; McVicker's Theatre bldg., 9 to 5; res. 102 Dearborn ave.

S

St. John, Leonard (R) McGill; Royal Col. Surg., England; Col. P. & S., Canada, 1873; 539 Monroe, 1 to 2, 7 to 8; tel. 7043. Prof. of Demonstrations of Surgery and Surg. Appliances, Col. P. & S., Chi.

Saalfeldt, J. (R) Royal Univ., Warsaw, 1846; 225 Larrabee; tel. 3049.

Sageser, Joseph S. (R) Jefferson, 1883; 329 Park ave., 12:30 to 2:30, 5:30 to 6:30; tel. 7048.

Saint-Cyr, Edelmar D. (R) Col. P. & S., Lower Canada, 1861; 379 Blue Island ave., 7 to 9 a. m., 1 to 4, 7 to 9 p. m.; tel. 9082.

Saint-Cyr, Emiline D. (R) Rush, 1888; 379 Blue Island ave., 7 to 9 a. m., 1 to 4, 7 to 9 p. m.; tel. 9082.

Salisbury, Jerome Henry (R) Rush, 1878; 175 Western ave., 8 to 10, 4 to 6; tel. 7263. Prof. Chemistry and Toxicology, Woman's Med. Col. and Northwestern Col. Dent. Surg.

Sandberg, Karl (R) Royal Univ. of Christiana, Norway, 1881; 223 W. Indiana, 1 to 3, 7; Sun., 12 to 1; tel. 4562.

Sanders, H. B. (H) Chi. Hom. 1882; 268 31st, 9 to 11, 7 to 8; res. 3245 Forest ave.

Sanders, W. H. (H) N. Y. Med. Col., 1866; 3245 Forest ave., 8 to 10, 2 to 4, 6 to 7; tel. 8488.

Sattler, Phil. (R) Rush, 1881; 422 W. 12th, 8 to 9, 12 to 3, 6 to 8; tel. 9189.

Saunder, Vida A. (H) Hahnemann, 1884; 70 State, 10 to 12, 3:30 to 4:30; res. 217 31st, 5 to 7:30; tel. 8108.

Saunier, A. J. C. (R) Univ. Mich., 1881; 181 W. Madison, 10 to 12, 7 to 9; tel. 4152. Attg. Phys. W. Side Free Dispensary; Lecturer on Histology and Microscopy, Col. P. & S.; Prof. Histology and Pathology, Chi. Ophthalmic Col.

Saur, Mrs. Prudence (R) Wom. Med., Phila., 1871; 192 Michigan ave. 10 to 3.

Sawyer, Edward Warren (R) Harvard Univ., 1873; 3733 Vincennes, till 9, 1 to 2, 6 to 7.

Schaefer, Frederick (R) Rush, 1873; 469 Milwaukee ave., 8 to 9, 1 to 2, 7 to 8; tel. 4639.

Schaefer, Fred. C. (R) Chi. Med. Col., 1876; 103 State, 1 to 2; res. 582 Washington boul., until 9, 4 to 5, 7 to 8; tel. 7103. Prof. Descrip. and Surg. Anatomy, Chi. Med. Col.

Schaefer, K. (H) Chi. Hom. Med., 1887; 156 Loomis, 8 to 10, 7 to 9.

Schaefer, Samuel (R) Univ. Leipzig, Germany, 1884; 255 North ave., till 9, 1 to 4, after 7; tel. 3366.

Schaller, Geo. J. (R) Rush, 1881; 582 Sedgwick; tel. 3607.

Schaller, J. (R) Heidelberg, Germany, 1854; 193 W. Randolph, 7 to 8, 12 to 2, 6 to 7:30.

Scheppers, Desire Q. (R) Rush, 1866; 292 Larrabee; res. 46 Bethoven pl.

Scheuer, Maurice (R) cor. 26th and Calumet.

Scheuermann, F. (H) Chi. Hom Med., 1879; 191 E. North ave., till 9, 2 to 4, 6 to 8; tel. 3152.

Schick, G. (R) Univ. N. Y., 1879; P. & S., 1887; 2215 Archer ave., till 9, 1 to 3, 6 to 8; tel. 8204.

Schirmer, Alfred (R) 547 Blue Island ave.; res., 514 S. Ashland ave.

Schirmer, Gustav (R) Erlangen, Bavaria, 1882; 547 Blue Island ave., 3 to 4; tel. 9085; res. 625 W. Taylor, 1 to 3; tel. 7178.

Schleyer, Chr. (R) Paris, L'Academic of Med., 1845; S. W. cor. Halsted and 14th, 8 to 10, 12 to 4, 6 to 8; tel. 9006.

Schloesser, A. G. (R) Rush, 1871; 78-80 State, room 32, 10 to 12, 3 to 6.

Schmidt, Ernest (R) Wurzburg, 1852; Rush, 1872; 78 State, 2:30 to 6; res. 1828 Wabash.

Schmidt, J. A. (R & H) Hahnemann, 1887; 477 26th, 8 to 9, 1 to 2, and nights; tel. 8372.

Schmidt, O. L. (R) Chi. Med., 1883; 78 State, 11 to 12:30: Alexian Brothers' Hospital, 2 to 3; res. 1828 Wabash ave.; tel. 8247.

Schneider, S. N. (H) Chi. Hom. Med., 1881; 302 Chicago Opera House, 1 to 3; tel. 1339; res. 210 Dearborn ave., 8 to 9, 6 to 7; tel. 3002.

Schottenfels, Emil (E) Bennett, 1874; 571 S. Halsted, 10 to 12; tel. 9006; 2427 Wentworth ave., 2 to 4; tel. 8099; res. 686 W. Adams, 7 to 10, 6 to 9.

Schrader, W. H. (H) Hahnemann, 1887; 3900 Cottage Grove, 9 to 11, 4 to 5:30, 6:30 to 7:30; tel. 9817; res. 3834 Johnson pl.

Schreiner, V. (R) Pesth, 1860; 302 North ave., 8 to 9, 2 to 3, 6 to 8; tel. 3208.

Shroeder, N. J. (R) Royal Bavarian Ludwig Maximilian Univ., Munich, 1860; 559 27th, till 9, 1 to 3, evenings; tel. 8372.

Schubert, J. J. (R) Rush, 1888; 355 S. Hoyne, 8 to 10, 5 to 7; tel. 7196.

Schultz, O. E. 2448 Cottage Grove ave.; tel. 8193.

Schwandt, E. H. (R) 1424 Indiana ave.; tel. 8438.

Schwuchow, Herman (H) 613 N. Ashland ave., 8 to 10 a. m.

Scott, T. A. (R) Rush, 1883; 783 W. 12th; till 9, 2 to 4, evening; tel. 7229.

Scroggy, G. H. Physio. Med. Inst., Chi., 1887; 1215 Van Buren, till 8:30, 12 to 1, 4:30 to 7:30; tel. 7206.

Scudder, H. Martin (R) Long Island Col. Hosp., N. Y., 1874; Argyle bldg., cor. Jackson and Michigan, 12 to 2; tel. 5570; res. 3446 Wabash ave., till 9, 7 to 8; tel. 8154. Phys. in Chief, Armour Mission Disp.

Sedgwick, Louisa (R) cor. Rhodes ave. and 32d; tel. 8353.

Seeley, Thaddeus P. (R) Union, Schenectady, N. Y., Michigan Univ., 1856; 289 W. Monroe, 8 to 9, 2 to 3, 6:30 to 7:30; tel. 4504.

Seiffert, Rudolph (R) Vienna, 1852; 171 and 173 E. Randolph, 2 to 4; tel. 1163.

Senier, Frederick S. (R) Rush, 1884; 245 Lincoln ave., 7 to 10, 1:30 to 2:30, after 6.

Seymour, Fred (R) Cincinnati Col., M. & S., 1856; 51 S. Halsted, 9 to 11, 1 to 4, 7 to 8.

Shaffer, Mary (R) Physio. Med. Col., Chic., 1887; 292 W. Madison, 1 to 4.

Shanahan, T. P. (R) Rush, 1877; 349 S. Clark, 8 to 11, 4:30 to 7; tel. 1401 and 1642.

Sharpe, James T. (R) Royal Col. Surg., Ireland, and King's and Queen's Col. Phys., 1865; 264 Ohio, 7 to 9 a. m., 11 to 3, after 6; tel. 3505.

Shaw, Siremba (R) Rush, 1882; 133 S. Halsted, 11 to 12, 5 to 6, 7 to 8; Sundays 10 to 12; tel. 4622.

Shaw, Thos. J. (R) Rush, 1879; 577 W. Congress, till 9, 12 to 2, and at night; tel. 7119. Clinical Asst. to chair of Gynæcology, Rush Med. Col.

Sheardown, T. W. (R) Jefferson, 1879; 143 Wabash ave, 8 to 5; res. 710 Pullman bldg.

Shears, Geo. F. (H) Hahnemann, 1880; 3130 Indiana ave., 8 to 9, 12 to 2, 6:30 to 7:30; tel. 8358. Prof. Surg., Hahnemann Col.; Supt., Hahnemann Hosp.

Shenick, O. T. (R) Rush, 1887; 62 S. Halsted, 10 to 11, 2 to 5; tel. 4152.

Sheppard, W. W. (R) P. & S., 1885; 161 W. Madison, 11 to 1; res. 209 Loomis, till 9, 1 to 3; tel. 4559.

Shepstone, J. A. (R) Bishop's Col., Montreal, 1881; 3818 State, 7 to 9, 2 to 4, 6 to 8; tel. 9824.

Sherman, Frederick E. (R) Rush, 1873; 203 Blue Island ave., 9 to 10, 2 to 4, 7 to 9; tel. 4160; Asst. Demonstrator of Anatomy, Rush Med. Col.

Sherry, Henry (H) Chi. Hom. Med., 1880; 3616 Stanton, 8 to 9, 1 to 2, 6 to 7. Surg. Cook Co. Hosp.

Sherwood, F. R. (R) Rush, 1888; 70 Madison; 7 p. m. till 10 a. m.; tel. 752; res. 356 37th, 11 to 12, 1 to 3, 4 to 6; tel. 9881.

Sherwood, W. T. (R) Univ. Mich, 1859; 1035 W. Van Buren, 9 to 12, 2 to 5; tel. 7280.

Shettler, G. O. (R) Julius Maximilian, Univ., Bavaria, 1856; Jefferson Eclectic Univ. 1863; 1602 Milwaukee ave., (Bandow P. O.)

Shimp, Archie J. (R) Jefferson, 1888; 196 E. Madison.
Shipman, Geo. E. P. & S., N. Y., 1843; 114 S. Wood; res. 120 S. Wood. Supt. Foundling's Home.
Shugart, Joseph (R) Ohio Med., 1864; 947 W. Harrison; res. 215 S. Lincoln.
Siegmund, Emilie (H) Chi. Hom. Med., 1881; 70 State, room 500, 10 to 12; res. 65 Wisconsin, 1 to 3; tel. 2587.
Silva, C. C. P. (R) Lisbon Col., Portugal, 1862; 163 State, 1 to 3; tel. 752; res. 98 Ogden ave., 10 to 12, 6 to 7:30; tel. 7172. Prof. Therapeutics Col. P. & S.
Simons, C. J. (R) Albany, 1867; 284 32d, till 8:30, 12 to 2, 6 to 7:30; tel. 8420.
Simpson, J. (R) Rush, 1867; 152 Oak, 9:30 to 1, 5 to 8; tel. 3036; res. 548 Cleveland ave., till 9, 2 to 4.
Simpson, W. L. 452 State.
Sincere, Emil (R) Louisville Univ., 1873; 82 Madison, 10 to 12; tel. 8114; res. 2974 Wabash, 1 to 3.
Sinclair, James G. (R) Bennett, 1883; Col. P. & S. Chi., 1888; 117 Center ave., till 9, 12 to 2, 6 to 7:30; tel. 3299; Asst. Phys. W. Side Free Disp.
Singley, Chas. C. (R) Jefferson, 1881; S. E. cor. State and Harrison, 8 to 11, 2 to 5, nights; tel. 1070.
Skeer, John D. (R) Cincinnati, 1852; Univ. Nashville, 1866; 681 Washington bv., 8 to 9, 12 to 1, 6 to 7; tel. 7203.
Skiles, Hugh P. (H) Iowa State Univ.; Hahnemann, 1880; 963 W. Monroe, 8 to 9, 1 to 3, 6:30 to 7:30 p. m.; tel. 7249.
Sloan, Henry H. (R) Chi. Med., 1869; 498 W. North ave., till 9:30, 1 to 3, after 7; tel. 4294.
Small, A. R. (R) Rush, 1874; 3300 State, 8 to 9, 12 to 1, 4 to 7; tel. 8100. Mem. Staff, S. Side Free Disp.
Small, H. N. (H) Hahnemann, 1866; 188 S. Clark, 12 to 3; res. 329 W. Van Buren, 9 to 10, 5 to 7; tel. 4311.
Smedley, N. J. (R) Chi. Med., 1887; 138 Wells, 8 to 10, 11 to 4, after 5; tel. 3624.
Smith A. K. 206 W. Lake, 8 to 9, 3 to 4, 6 to 9; res. 161 Walnut; tel. 4603.
Smith, A. W. (E) Eclec. Med. Inst., Cincinnati, 1872; 1012 W. Lake; res. 1549 W. Monroe.
Smith, Charles Gilman (R) Univ. Penn. 1851; N. E. cor. State and Madison, 12 to 2, 3 to 4, Sun. 12:30 to 2; res. 2220 Calumet ave., 8 to 9 a. m., 7 to 8.
Smith, David S. (H) Jefferson Med. Col, 1836; 1255 Michigan ave. Emeritus Prof. of Materia Medica and Therapeutics, Hahnemann Col. Chi.
Smith, Edwin H. (R) McGill, 1884; 2642 Wentworth ave., till 9, 12 to 2, 5 to 6, 7:30 to 8:30.
Smith, Espey L. (H) Chi. Hom., 1883; 974 W. Polk; till 9, 12:30 to 2:30, 6 to 8:30; tel. 7253.
Smith, E. M. (R) Chi. Med., 1886; 626 W. Lake, 8 to 10, 1 to 3, 7 to 8; tel. 7015.

Smith, Jennie E. (H) Chi. Hom. Med, 1879; 665 Sedgwick; tel. 3109.

Smith, Julia Holmes (H) Chi. Hom., Med., 1877, 521 Dearborn ave. 9 to 12, 6 to 7:30; tel. 3310.

Smith, W. H. F. (R) Rush, 1884; 214 Clark; res. Windsor Hotel.

Smith, Wm. F. (R) Miami Med. Col., Cincinnati, 1868; 204 Dearborn, 11 to 2.

Snyder, A. F. (R) Rush, 1884; 1251 Madison, 9 to 11, 1 to 3, 7 to 8; tel. 7007; res. 1205 Monroe, till 9, 12 to 1, 6 to 7; tel. 7007; Attg. Phys., Central Free Dispensary.

Snyder, Omer C. (H) Chi. Hom. Med., 1884; 190 Cass, 8 to 10, 3 to 5; tel. 3125.

Snyder, O. W. F. Physio. Med. Inst., Cincinnati, 1878; 125 Clark, 10 to 4; tel. 1339.

Spalding, Heman (R) Chi. Med. Col., 1881; 235 State, 10 to 12, 1 to 3, night; tel. 2349.

Sperry, Chas. C. (R) Chi. Med., 1874; 3100 Archer ave., 10 to 11, with Union Steel Co., 8 to 9, at 3; tel. 9605; res. 3243 S. Paulina.

Stahl, E. L., Jr., (R) Rush, 1882; 171 E. Van Buren, till 10, 1 to 3, after 7.

Stanley, C. W. (R) N. Y. Univ., 1866; 2113 State, 2 to 3, 7 to 9.

Stanley, F. A. (R) P. & S., N. Y., 1870; 90 S. Morgan, 7 to 11, 1 to 3, evening; tel. 4504.

Stansbury, Mrs. H. E. (H) Chi. Hom. Med., 1877; 343 W. Monroe, 10 to 12, 3 to 5; tel. 4062. Supt., Talcott Nursery.

Stanton, J. F. (R) P. & S., 1888; 395 W. Harrison, 9 to 11, 2 to 4, 7 to 9; tel. 4637; res. 182 N. Curtis, till 9, 12 to 5, after 9; tel. 4562; Attg. Phys., W. Side Free Disp.

Stanton, Jas. T. (R) P. & S., 1888; 395 W. Harrison, 9 to 11, 2 to 4, 7 to 9; tel. 4637; res., 182 N. Curtis, till 9, 5. to 7, after 9 tel. 4562. Asst. Phys. W. Side Free Disp.

Starkey, Horace M. (R) Chi. Med. Col., 1878; 125 22d, 3 to 4:30; tel. 8184; res. 3302 Indiana ave., 8 to 9, 6 to 7, Sunday, 9 to 10; tel. 8479; Att'g. Phys. Mercy Hospital; Att'g. Surg. Eye and Ear Dept. S. Side Free Disp.

Stearns, W. M. (H) Chi. Hom. Med., 1880; 100 State, 1 to 5; res. 15 Warren ave.; Ass't to Chair Eye and Ear, Chi. Hom. Med. Col.; Ass't Eye and Ear Surg. Cen. Hom. Hosp. and Free Disp.

Stebalts, Frank W. (A) Vienna, Austria, 1868; 729 S. Halsted, 10 to 11, 4 to 5; tel. 9031; res. 323 S. Halsted, till 8:30, 1 to 3, night; tel. 4088.

Steele, D. A. K. (R) Chi. Med., 1873; 1801 State, 9 to 10, 3 to 5; tel. 8202; res. 2920 Indiana ave., till 9, after 6; tel. 8220; Prof. Orthopedic Surg. Col. P. & S., Chi.; Att'g. Surg. Cook Co. Hosp.

Steger, R. W. (R) P. & S., N. Y., 1878 ; 7 Lakeside Bldg., 11 to 3:30 ; tel. 1339 ; res. Hotel St. Benedict, 355 Chicago ave., till 9, after 5; tel. 3125.

Stehman, H. B. (R) Jefferson, 1877 ; 300 S. Wood, till 1, after 4 ; tel. 7189; Med. Supt. Presb. Hosp.

Steinhaus, H. (H) Hahnemann, 1878; 479 Noble, 8 to 10, 6 to 8.

Stephani, Alfred H. (R) Rush, 1884 ; 156 W. Randolph; tel. 4614 ; res. 904 W. Madison.

Sterl, A. (R) Univ. N. Y., 1866; 361 Blue Island ave.

Stern, D. H. (E) N. Y. Eclectic, 1882; 202 E. Chicago, 10 to 12, 2 to 6, after 7; tel. 3274; res. 182 E. Indiana, 7:30 to 10, 1 to 2.

Steurnagel, G. (R) Michigan Col. Med., 1883; 2876 Archer ave., 8 to 10, 2 to 4, 7 to 9; res. 3120 Lowe ave., 11 to 2, 5 to 7, nights; tel. 8274.

Stevens H. M. (H) Chi. Hom., 1885; 1347 W. Madison, till 9, 1 to 2, 6 to 7; tel. 7118.

Stevenson, Sarah Hackett (R) Wom. Med., 1875; Central Music Hall, 10 to 12, 5 to 6; tel. 5642; res. Dearborn ave. and Erie. Prof. of Diseases of Women, Chi. Training School; Med. Dept. Chi. Hosp. for Women and Children; Consulting Phys. Woman's Hosp. and Erring Women's Refuge; Prof. Obstetrics, Woman's Med. Col.

Stewart, Andrew, (R) McGill, 1883; Royal London, 1884; 329 W. Van Buren, till 9, 1 to 2, 7 to 8.

Stillians, D. C. (R) Chi. Med., 1869; 58 State, 2 to 4; tel. 5558; at 480 Milwaukee ave., 8 to 9, 6 to 7.

Stockham, Alice B. (H) Chi. Homeo. Col.; 161 LaSalle; res. 6058 LaSalle, Englewood; tel. 209.

Stockwell, J. S. (R) Rush; 144 Oakwood bv., 9 to 11, 2 to 3, 7 to 8; tel. 9966; res. 3560 Vincennes; tel. 9804.

Storck, C. (R) Albany, 1865; 153 Dearborn ave.

Storer, W. D. (R) 26th and Calumet; tel. 8267.

Storey, Chas. A. (R) Wooster Univ., 1882; 52 31st, 10 to 12; tel. 8354; res. 3001 Vernon ave., 2 to 4, 7 to 8; tel. 8544.

Stout, Alexander M. (R) Univ. Georgetown, D. C., 1880; 339 S. Jefferson, 9 a. m., 7 p. m.; tel. 4593; 1123 Harrison, till 9, 2 to 6; tel. 7214.

Stowell, James H. (R) Chi. Med., 1881; 526 Wabash, 8 to 9, 3 to 5, 6 to 7; tel. 440.

Stratford, Henry Knox (E) Eclectic Med. Col., Philadelphia; 243 State, 11 to 5; tel. 309; res. Austin, 7 to 9:30 a. m., 7 to 9 p. m.

Streeter, John W. (H) Hahnemann, 1868; 2001 Prairie ave., 8 to 9, 12 to 2, 6 to 7. Prof. Med. and Surg. Diseases of Women, Chi. Homeo. Med. Col.; Gynæcological Dept., Central Homeo. Hospital and Free Disp.; Gynæcologist, Cook Co. Hosp.

Strehz, Theodore, (E) 380 Wells, 8 to 10, 2 to 5, after 7.

Stringfield, C. Pruyn 2448 Cottage Grove, 9 a. m. to 9 p. m.; tel. 8193.

Stringfield, F. M. (R) Georgetown Col., 1870; 2448 Cottage Grove ave., till 9, 4:30 to 7; tel. 8193.

Strong, Albert B. (R) Rush, 1872; 312 W. Indiana, 8 to 9, 1 to 2, 5 to 6, 7 to 8, Sun. 10 to 12; tel. 4235; res. 533 W. Monroe, till 8, 12 to 1, 6 to 7; tel. 7144.

Strong, Edward D. (E) Bennett, 1884; 47 N. Ashland.

Struble, John (E) Bennett, 1879; Western and Armitage, till 10 a. m., after 3; tel. 4210, P. O. Bandow, Ill.

Struble, J. R. (E) Bennett, 1886; 908 California ave., till 10, after 3; tel. 4252.

Struh, Carl (R) Univ. Zurich, Switzerland, 1886; 621 W. 12th, 8 to 10, 2 to 4, 7 to 8; tel. 7011.

Strzyzowski, Wladislaw (R) 723 W. 18th, 10 to 12, 2 to 4; tel. 9078; res., 683 W. 18th, till 10, after 5.

Stubbs, J. E. (R) Univ. Penn., 1864; 126 State, 8 to 9:30; 2 to 4; tel. 624; res. 68 E. 18th, after 6, till 8 a. m.; tel. 8045.

Sullivan, Daniel H. (R) Chi. Med., 1881; 2495 Archer ave., 8 to 10, 2 to 4, 8 to 9; tel. 8333; res. 2512 Archer ave.

Sumney, J. (R) Jefferson, 1864; S. E. cor. Clark and Van Buren, 8 a. m. to 8 p. m.; tel. 1155; res. 3837 Indiana ave.; tel. 9942.

Swain, U. J. (R) Albany, 1866; 262 W. Randolph, 9 to 10, 4 to 5, 8 to 9; res. 43 S. Elizabeth, till 9, 12 to 2, 6 to 8.

Swartz, T. B. (R) 1434 Indiana ave.

Sweet, E. C. (H) Univ. Mich., 1868; Am. Eclectic Col., 1870; Hahnemann, Chi., 1884; 70 State, rooms 407–408, 12:30 to 3; 557 W. Madison, 8 to 9, 6:30 to 7:30; tel. 7172.

Synon, Geo. C. (R) Rush, 1880; 249 Blue Island ave., 11 to 1, 4 to 6; tel. 4610; res. 462 W. Taylor.

Synon, W. A. (R) Rush, 1882; 249 Blue Island ave., 9 to 11, 2 to 4; tel. 4610.

T

Tabor, F. S. (R) Rush, 1881; 70 Madison, 10 to 4; tel. 753; res. 3523 Ellis ave.; tel. 9804.

Tagert, Alonzo D. (R) Vt. Univ., 1864; 966 W. Lake, until 9, 12 to 1 and 6 to 7; tel. 7142.

Tagert, A. H. (R) Vt. Univ., 1866; 966 W. Lake, 8 to 9, 1 to 2, 7 to 8; tel. 7142; res. 853 Walnut.

Talbot, Eugene S. (R) Penn. Col. Den. Surgery, 1873; Rush Med. Col. 1882; 125 State, 9 to 5. Prof. Den. Surg., Woman's Med. Col.

Talcott, J. B. (H) Western Reserve, 1845; 79½ 22d, 8 to 10, 1 to 4, 7 to 8; tel. 8133; res. 2112 Michigan ave.

Taliaferro, Frank (R) Jefferson, 1875; N.W. cor. Center ave. and 14th, 9 to 10, 2 to 3, 7 to 8; tel. 9081; res. 295 Center ave., till 8:30, 12:30 to 1:30, 6 to 7; tel. 4035.

Tascher, John (E) Eclec. Med. Inst., Cincinnati, 1878; Bennett Med. Col., 1880; 518 W. Chicago ave, 7 to 8, 1 to 2, 7 to 8; at Division and Milwaukee ave., 10 to 11; tel. 7170. Prof. Diseases of Children, Bennett Col.

Taylor, Charles S. (R) Chi. Med., 1875; 481 Ogden ave., 10:30 to 2; tel. 7177; res. 433 S. Oakley ave., till 8:30, and evenings.

Taylor, E. B. (E) Bennett, 1877; 243 State, 8:30 to 11:30.

Taylor, Geo. O. (R) Rush, 1868; 205 La Salle, 10 to 4; tel. 1348; res. 1339 Oakwood boul.

Taylor, T. T. (R) Ky. School Med., 1869; 186 W. Madison; 83 S. Halsted,

Teare, John (R) Glasgow, 1842; Royal Col. Surg., 1842; 411 Center, 12 to 2; tel. 3345.

Tebbetts, F. M. (R) P. & S., 1885; cor. Halsted and Polk; tel. 4388; 12th and Morgan; tel. 4771.

Terry, Junius (R) St. Louis, 1861; 378 Wabash ave., 10 to 2.

Thacher, C. I. (H) Homeo. Hosp. Col., Cleveland, 1880; 6 Central Music Hall, 9 to 6; res. 51 Aberdeen, 7 to 9, 6 to 9.

Thackeray, W. T. (R) Jefferson, 1866; 163 State, 11 to 4; res. 341 N. Clark, till 10, 5 to 7.

Theobald, G. (H) Chi. Hom. Med., 1883; 750 S. Halsted, 7:30 to 10, 5 to 8; tel. 9021; res. 847 S. Ashland ave.

Thies, Wilhelm (R) Jefferson, 1877; Friedrich Wilhelm Univ., Berlin, 1882; 156 W. Randolph, 3 to 5; tel. 4614; res. 136 Canalport ave., till 9, 1 to 3, after 6; tel. 9062.

Thilo, G. (R) Strasburg, Germany, 1877; 493 Milwaukee ave., 8 to 10, 1 to 3, after 7.

Thomas, A. L. (R) Chi. Med., 1879; 3100 Wentworth av., till 9, 12 to 2, 7 to 9; tel. 8465.

Thomas, G. S. (R) Albany, 1858; 333 W. Adams.

Thomas, Homer M. (R) Rush, 1882; 34 Throop, 8 to 9, 12 to 1:30, 5 to 7:30; tel. 4077. Lecturer Diseases Throat and Chest, Woman's Medical Col.; Phys. and Surg. for Diseases Throat and Chest, Cent. Free Disp.

Thome, Arthur G. (H) Chi. Hom. Med., 1883; 239 Lincoln av., till 9, 1 to 2, 6 to 8; tel. 3404.

Thometz, J. J. (R) Rush, 1882; 564 S. Halsted, 9:30 to 10:30, 2 to 3; tel. 9030; res. 999 W. 12th, till 9, after 6; tel. 7162.

Thompson, Geo. A. (R) Univ. of N. Y., 1884; 52 31st, 10 to 12, 4 to 6, 7 to 8; tel. 8354; res. 79 31st. Attg. Phys. Dept. of Laryncology, W. Side Free Disp.

Thompson, J. J. (H) Chi. Hom. Med., 1888; 251 Hermitage av., 8 to 10, 1 to 2, 6 to 8; tel. 7134.

Thompson, Mary Harris (R) Chi. Med. Col., 1870; 26 Cent. Music Hall, 10 to 12:30; res. 638 W. Jackson; tel. 7037. Prof. Gynæcology, Hosp. Women and Children.

Thompson, M. M.(H) Chi. Hom. Med., 1886; 403 Oakley av., 8 to 9, 1 to 2, 6 to 7, Sun. 12 to 2; tel. 7006.

Thompson, Merritt W. (R) Rush, 1877; 282 W. Indiana, before 9, 1 to 2, 7 to 8, nights; Sun. 12 to 1; tel. 4254.

Thornton, Frances E. (E) Bennett, 1888; 108 Washington, 9 to 11; tel. 1339; res. Avondale, Ill.; 7 to 8:30, 1 to 7.

Thuemmler, Alexander (R) St. Louis Med. Col., 1879; cor. Ashland and Milwaukee avs., 8 to 9; res. 276 Divsion, 12 to 1; tel. 7261.

Thurston, Ebenezer H. (R) Buffalo, 1865; 518 Wabash, to 12, 2 to 5, 8 to 10; tel. 8451; res. 3018 Indiana av., 5 to 7, till 9.

Tilley, R. (R) Chi. Med., 1876; 125 State, room 17, 9 to 1; res 1436 Michigan av.

Tobias, G. Jackson (R) P. & S., 1885; 247 W. Madison, 10 to 1, 4 to 6, Sat. eve.; tel. 4310; res. 173 S. Western av.; tel. 7263.

Todd, James F. (R) Bellevue, 1863; 171 E. 22d; 10 to 12, 2 to 4, 7 to 8; tel. 8165; res. 2447 Prairie av.

Tomboeken, Henry (R) Rush, 1866; 171 E. Madison; 9 to 12, 2 to 3.

Tomlinson, William M. (R) Univ. Pa., 1849; Health office, 9 to 4; tel. 447; res. 583 Adams, till 8, after 5, Sun. 3 to 6.

Tongue, F. J. (E) Bennett, 1878; 210½ Clark.

Tons, E. L. (R) Ind. Med. Col., 1874; 694 W. Van Buren, 8 to 12, 1 to 5.

Tooker, Robt. N. (H) Bellevue, 1865; 261 Dearborn ave., till 10, 3 to 5; tel. 3278. Prof. Diseases Children, Chi. Hom. Med. Col.

Traub, Christopher (H) 749 N. Paulina, 9 to 10, 2 to 4; tel. 4774.

Treat, R. B. Eclec. Med. Inst., Cincinnati, 1847; Berkshire, 1866; 247 W. Madison, 9 to 11, 2 to 4, eve.; tel. 4310.

Trimble, E. G. (R) Col. P. & S., N. Y., 1878; 84 Lincoln ave., 8 to 10, 5 to 7.

Trine, John G. (H) N. Y. Hygieo-Ther. Col., 1859; 45 E. Randolph, till 9 to 1, 2 to 6; res. 3621 Ellis Park.

Tripp, Robinson 1408 Wabash ave.

Trout, Elizabeth H. (R) Wom. Med., 1884; 880 W. Jackson, till 9, 1 to 3, 6 to 7; tel. 7159.

Tucker, D. Mills (R) Harvard, 1852; 441 State, 10 to 11, 2 to 3, 8 to 9; res. 2317 Wabash ave.

Tucker, H. S. (E) Bennett, 1879; 513 State, 10 to 12, 1 to 4, Mon., Wed. and Fri. 6 to 8; tel. 667; res. 464 42d; tel. 9892. Prof. Descriptive Anatomy, Bennett Col.

Tucker, James I. (R) Harvard Univ., 1867; 52 35th, till 9, 1 to 3, 6 to 7; tel. 9804.

Tully, A. Melville (R) Univ. Vt., 1880; Pullman bldg., 12 to 1; 205 N. State, till 11, 3 to 5; tel. 3125.

Turnock, Edwin (R) Chi. P. & S., 1886; 109 N. Clark; tel. 3158.

Tuthill, J. J. (R) Rush, 1883; 858 W. Van Buren, 8 to 9, 1 to 2, 5 to 6, 7 to 8; tel. 7159; res. 297 S. Oakley.

Twining, S. Douglas (R) Yale, 1864; 210 W. Indiana, till 9, 2 to 3, 7 to 8; tel. 4562.

U

Ulrich, Julius (H) Univ. Vienna, 1847; 202 Center ave., till 10, 4 to 8; tel. 3407.

Ulrich, R. (R) Univ. of Wurzburg, 1871; 202 Center ave., till 9, after 5; tel. 3407.

V

Valentine, J. C. Hahnemann, 1886; 1332 Wabash, 10 to 12, 1 to 3, 6:30 to 7:30.

Valentine, Sarah L. (H) Hahnemann, 1886; 1332 Wabash, 11 to 12, 2 to 4.

Valpey, J. W. (R) P. & S., N. Y., 1857; 544 Blue Island ave., 9 to 10, 1 to 3, after 7.

Vampill, R. (R) Maryland Univ.; 1857; 405 Wells, 10 to 12, 2 to 4.

Van Buren, Henry (R) Rush, 1865; Washington bvd. and May, 8 to 10, 5 to 7; tel. 4656; res. 15 N. May.

Van Doozer, B. Rel. (R) Chi. Med., 1868; 3201 Calumet ave., till 9, 2 to 3.

Van Hook, Weller (R) P. & S., 1885; 884 W. Madison, 8 to 9, 2 to 3, after 6:30; tel. 7176.

Van Hook, W. R. (R) Univ. Louisville, 1860; 131 Sacramento ave., 7 to 9; 12 to 2, after 7; tel; 7118.

Vary, W. H. (H) Hahnemann, 1888; 50 Walnut, 8 to 11, 2 to 4, 7 to 9.

Venn, Chas. H. (R) Rush, 1876; 547 Milwaukee ave.

Venn, Clement (R) Rush, 1887; 167 N. Clark, till 10, 1 to 3, after 7; tel. 3166.

Venn, Henry (R) Royal P. & S., Ontario, 1861; 99 Blue Island ave., 8 to 10, 1 to 3, 6 to 8; tel. 4101.

Verity, W. P. (R) Rush, 1879; 62 E. Chicago ave., till 11, after 8.

Vilas, Charles H. (H) Hahnemann, 1873; Central Music Hall, 9 to 1; res. Union Club. Prof. of Diseases Eye and Ear, Hahnemann Med. Col.; Senior Surg. Eye and Ear Dept. Hahnemann Hosp; tel. 2642.

Vincent, Mary L. (R) Univ. Mich., 1875; 70 State, 10 to 1; res. 144 22d.

W

Wadsworth, F. L. (R) Rush, 1869 ; 607 Division, 1 to 3 ; tel. 3070. Prof. Physiology Wom. Med. Col.

Wagner, C. B. (R) Chi. Med., 1888; Michael Reese Hosp.

Waldmeyer, Jos. R. (R) State Med. Soc., N. J., 1875 ; 82 W. Madison, 11 to 12, 1 to 4; tel. 4753.

Walker, Chas. E. (R) Rush, 1885 ; 85 Washington, 9 to 11 ; res. 321 Park ave.; tel. 1339.

Walker, Geo. B. (R) Nat. Eclec. Med. Col., 1870; Eclec. Med. Col. of Penn., 1877; 125 S. Clark, 1 to 4 ; res. 38 Walnut, till 9, 5 to 7; tel. 7016.

Walker, J. B. 85 Washington, 10 to 4; res. 321 Park ave.

Walker, J. Warren (R) Rush, 1884; 894 W. Lake, till 7, 10 to 12, 3 to 5, 7 to 9.

Walker, Sydney (R) Nashville Med. Col., 1884; 193 S. Clark, 10 to 4; res. 419 W. Monroe.

Wallace, Geo. B. (R) Ind. Med. Col., 1877; 420 S. Clark, 9 to 11, 3 to 6, 7 to 9; tel. 1690; 347 5th ave., 1 to 3.

Ward, Charles W. (R) Rush, 1887 ; Pres. Hosp.; tel. 7189 ; Senior Interne, Pres. Hosp.

Ware, Lyman (R) Chi. Med., 1866 ; Univ. Penn., 1868; 125 State, 9 to 11, 4 to 5; tel. 8233; res. 1620 Prairie ave. Oculist and Aurist, Pres. Hosp.; Oculist Chi. Orphan Asylum ; Surg. Eye Dept., Ill. Char. Eye and Ear Inf.

Washburne, Geo. F. (H) Chi. Hom. Med., 1885; 9 Warren ave., 9 to 10, 5 to 7; tel. 7102.

Wassall, J. W. (R) Col. P. & S., Chi., 1884 ; 208 Dearborn ave., 9 to 5; tel. 3125 ; Dentistry. Prof. of Regional Anatomy and Physiology in Chi. Col. of Dental Surg.

Waters, Frank R. (H) Hahnemann, 1888; 1735 State, 8 to 10, 2 to 4, 7 to 9.

Waterous, Daniel S. (R) 170 Evergreen av.

Watry, Joseph (H) Hahnemann Med. Col., 1883 ; Central Music Hall, 9 to 12:30; tel. 5642; res. 1668 Wellington ave., Lake View. Clinical Prof. of the Diseases of the Eye and Ear in Hahnemann Hosp.; Adj. Prof. to Eye and Ear Chair, Hahnemann.

Watson, Lewis H. (R) Cincinnati Med. Col., 1866; 297 Indiana, till 9, 1 to 2, 4 to 6; tel. 3419.

Way, James P. (R) P. & S., 1886; 336 W. 12th, 8 to 10, 7 to 9 p.m., night; tel. 4061.

Waxham, Frank E. (R) Chi. Med., 1878; 70 Monroe, 1 to 3; tel. 80; res. 3449 Indiana ave., 8 to 9, 6 to 7; tel. 8351. Clinical Prof. Laryngology and Rhinology, Chi. Ophthalmic Col.; Prof. Diseases of Children, Col. of P. & S.

Weaver, William H. (R) P. & S., 1883; 394 Belden ave., till 10, 12 to 3, 7 to 8; tel. 3317. Phys. to Home for Incurables and Gynæcological Dept., Chi. Policlinic Disp.

Webb, Frank R. (R) Chi. Med. 1875; 2894 Archer ave., 9 to 11, 7 to 9; res. 3907 Michigan ave.; tel. 8297.

Webb, W. J. (R) Univ. of Mich., 1878; 70 State, 12 to 2; tel. 5558; res. 360 E. Division, till 10, after 6; tel. 3284.

Webster, C. E. (R) Harvard, 1883; Opera House block, 10 to 2; res. 104 Oak, till 9, 6 to 8.

Webster, George W. (R) Chi. Med., 1882; 505 S. Canal, 9 to 10, 4 to 5, 7 to 8; Sun., 10 to 11; tel. 4494; res. 1922 Indiana ave., 12 to 2; tel. 8044. Lecturer Physiology, Chi. Med. Col.

Webster, John C. (R) Harvard, 1867; 820 W. Jackson, 8 to 10, 1 to 2, 5 to 7; tel. 7188.

Weeks, Geo. H. 1 Central Music Hall, 11 to 1; res. 2500 Indiana ave.; tel. 8122

Weeks, J. A. Central Music Hall.

Wegner, O. (H) Chi. Hom. Med., 1882; 875 Milwaukee ave., 8 to 10, 2 to 4, 6 to 8; Sun., 9 to 10, 2 to 3; res. 677 Leavitt; tel. 4118.

Weilhart, Mrs. C. E. (H) Chi. Hom. Med., 1871; 3709 Ellis ave., 9 to 10, 2 to 4.

Weller, F. Montrose (R) Univ. Mich., 1854; 159 Clark, 12 to 3; res. 243 31st.

Wells, Catharine J. (H) Hahnemann, 1883; 70 State, 10 to 12, 3:30 to 4:30; res. 217 31st, 5 to 7:30; tel. 8108.

Wescott, Austin A. (E) Eclec. Med. Inst., Cincinnati, 1857; 57 S. Elizabeth, till 9, 1 to 2, 5 to 7; tel. 4638.

Wescott, Cassius D. (R) Rush, 1883; cor. Madison and Ogden ave., 10 to 12, 6 to 7; tel. 7172; res. 56 S. Elizabeth, at night; tel. 4638. Lecturer on Anatomy, Rush Med. Col.

Wesley, A. A. (R) Chi. Med., 1887; 31st and State, 9 to 10:30, 4 to 5, 7 to 8; res. 2952 State; tel. 8166.

Westerburg, Richard (R) Wurzburg, 1866; 783 Milwaukee ave., till 10, 2 to 4, eve.; tel. 4573.

Weston, Edward B. (R) Rush, 1873; 65 Randolph, 2:30 to 4; tel. 5558; res. 3225 Vernon ave., 6 to 7; tel. 8602.

Wetherla, W. W. (R) Chi. Med., 1885; 315 North ave., till 9:30, 11 to 1, 3 to 5, 7 to 9, nights; tel. 3208.

Wetherell, Geo. F. (R) Univ. N. Y., 1856; 113 Adams, 10 to 3; res., 164 Oakwood boul., 8 to 9, 5 to 6.

Whaling, Julia Cone (H) N. Y. Hom. Col., 1859; Clifton House, 10 to 3

Wheeler, F. A. (E) Bennett, 1885; 581 State, 10 to 4.

Whidden, Philon C. (R) Harvard Univ., 1866; 125 S. Western ave., 8 to 9, 12 to 2, 6 to 8; tel. 7057.

White, Carrie Noble (R) Wom. Med., 1887; 358 Ogden ave., 10 to 1, 4:30 to 6; tel. 7196.

White, Chas. (R) Rush, 1863; 380 LaSalle ave., 9 to 10, 2 to 4, evening.

White, G. J. (R) P. & S., 1887; 119 Madison, 10 to 1; tel. 1339.

White, George (R) Kentucky School of Med., 1881; 1251 W. Madison, 9 to 10, 1 to 2, 7 to 8; tel. 1007.

White, H. A. (E) Eclectic Med. Inst., Cincinnati, 1882; 886 W. Madison, 10 to 2, 6 to 9; tel. 7176.

White, J. E. (R) Rush, 1883; 358 Ogden ave., 8 to 10, evenings; tel. 7196.

Whitfield, G. W. (R) Chi. Med., 1887; 240 Wabash, 10 to 1, 3 to 5, evening; tel. 2457.

Whitford, H. K. (E) Eclec. Med. Inst., Cincinnati, 1861; 511 State; Tues. and Fri., 9 to 4; tel. 667; res. Elgin, Ill., tel. 21 and 23. Emeritus Prof. of Principlel and Practice of Med., and Prof. of Clinicas Med., Bennett Med. Col.; Prof. Gen. Diseases, Bennett Free Disp.

Whiting, Harry C. (R) Chi. Med., 1888; 3021 State, 10 to 2, 7 to 8; tel. 8126.

Whitman, Charles H. (E) Bennett, 1886; 119 Madison; res., 5965 Wentworth ave.

Whitney, Eugene W. (R) Rush, 1878; S. E. cor. Van Buren and Halsted, till 10, 12 to 1:30, evening; Sun. till 10; tel. 4640. Demonstrator of Anatomy and Lecturer on Surgery, Rush Med. Col.

Whitney, L. W. (R) Western Reserve, Cleveland, 1877; 537 Lake, till 8, 6 to 9; tel. 7241.

Wickersham, Swayne (R) Univ. Pa., 1855; 235 State, 9 to 10, 3 to 4, 7 to 8.

Wiggin, T. B. (R) P. & S., 1886; 70 Monroe, 11 to 1; tel. 2480; res. 79 31st, 9 to 10, 3 to 4, 7 to 8; tel. 8354.

Wight, Eli (E) Bennett, 1885; 108 Washington, 12 to 1; tel. 1294; res. 856 Warren ave., till 10, 5 to 7; tel. 7118. Prof. Obstetrics, Bennett Col.

Wilbur, C. A. (H) Castleton Med. Col., 1852; 355 N. Clark, 8 to 10, 5 to 6.

Wilcox, Colin H. (R) Rush, 1888; 1339 W. Madison

Wild, Theo. (R) Rush, 1865; 465 Milwaukee ave., 2 to 4, 8 to 9; tel. 4575; res. 697 N. Robey, till 9.

Wilder, Flavius M. (R) Univ. Mich., 1868; 2515 Wabash ave., 8 to 9, 2 to 3, 9 to 7; tel. 8335. Throat and Chest Dept. S. Side Free Disp.

Wildman, H. G. 224 State; res. 348 Michigan ave.

Wilke, William M. (H) Chi. Hom. Med., 1878; 273 N. May, 1 to 3; tel. 4029.

Wilkin, J. S. (R) Buffalo, 1862; 225 Dearborn, 10 to 4; res. 764 Walnut, till 9, 6 to 9 p. m.

Wilkins, J. R. (R) Starling Med. Col., Columbus, 1856; 35 and 36 Ashland block, 10 to 4; res. 155 S. Hoyne ave.; tel. 5453,

Willard, Alfred L. (E) Bennett, 1878; 218 S. State, 9 to 5. Prof. Osteology and Morbid Anatomy, Bennett Med. Col.; Prof. of Pathology, Chi. Col. of Ophthalmology and Otology.

Willard, Geo. E. (R) Chi. Med., 1874; 3169 Archer ave., 3 to 5 ; res. 3342 S. Paulina, till 9, 6 to 9.

Williams, C. A. (H) Hahnemann, 1861 ; 103 State, 11:30 to 1 ; tel. 7139; res. 14 Warren ave.

Williams, D. H. (R) Chi. Med., 1883 ; 3034 Michigan ave., 9 to 10 3:30 to 5, and nights.

Williams, Edwin C. (H) Chi. Hom. Med., 1886 ; 3214 Graves pl., 8 to 9, 1 to 3, 6:30 to 7:30; tel. 8084.

Williams, J. B. (R) Univ. Penn., 1881; 358 Ogden ave., 7 to 9 a. m., 12 to 1, 6 to 9; tel. 7196.

Williams, John F. (R) Chi. Med., 1865 ; 427 Center, 10 to 1 ; tel. 3072.

Williams, L. R. (R) Rush, 1875; 68 Randolph, 9 a. m. to 8 p. m. ; res. 629 Sedgwick.

Williamson, J. G. (E) Bennett, 1882 ; 125 S. Clark, 11 to 3 ; res. Hinsdale.

Wilms, Edwin O. (R) Miami Col., Cincinnati, 1886; Clark and Harrison, 2 to 5; res. 124 Fullerton ave., 8 to 9, 7 to 8; tel. 3648.

Wilson, John H. (R) Harvard, 1881; 220 Dearborn ave., 7 to 10, 4 to 5, 6 to 8; tel. 3002. Phys. to Home for Incurables; Neurologist to Chi. Policlinic.

Wimermark, Arvid H. (R) Rush, 1884; 68 E. Chicago ave., till 9, 1 to 2, 7 to 8; tel. 3113. Asst. Surg. Chi. Policlinic Hosp., and Surg., to Chi. Policlinic Disp., Gynæcological Dept.

Windrow, Sven (R) Carolina M. & S. Inst., Stockholm, 1881; Univ. Penn., 1887; 64 E. Chicago ave., till 10, 2 to 4, after 6; tel. 3113. Asst. Eye Dept. Chi. Policlinic.

Winer, J. K. (R) Rush, 1884; 205 E. Ohio, 8 to 10, 1 to 3, 6 to 8; tel. 3045.

Wing, Elbert (R) Rush, 1882; 3266 Cottage Grove ave., 9 to 11, 3 to 5, 7 to 8 ; tel. 8354. Dem. Pathology, Chi. Med. Col.

Wolford, W. H. (E) Bennett, 1885; 3904 State, 8 to 10, 4 to 9; tel. 9853 ; res. 3732 State, till 8, after 9 ; tel. 9869.

Wolgamott, Geo. W. (R) Keokuk, Iowa, 1861; 70 State, 10 to 4; tel. 2558; res. 47 S. Elizabeth, till 9, after 5 ; tel. 4090

Woodbury, W. H. (H) Hahnemann, 1866; Central Music Hall, 11 to 12, 2 to 4; res. 315 Washington boul., 7 to 9, 7 to 8; tel. 4583.

Woodward, A. W. (H) Hahnemann, 1865 ; 130 Ashland ave. ; tel. 7102. Prof. Materia Med. and Therapeutics, Chi. Hom. Med. Col.

Woodworth, P. M. (R) Chi. Med. Col., 1878 ; 411 Center, till 8, 1 to 2; tel. 3345.

Wright, F. R. (R) Rush, 1885; 3714 Ellis ave., 9 to 11, 2 to 4; tel. 9874.

Y

Yates, George F. (R) Rush, 1888; 1237 W. Jackson, till 9, 1 to 2, evening; tel. 7206.

Young, J. (H) Hahnemann Med. Col., 1888; 81 29th, 9 to 10, 2 to 3.

Z

Zeisler, Joseph (R) Univ. of Vienna, 1882; 125 State, 10 to 2; tel. 5442; res. 2924 Groveland; tel. 8210.

A. NIEHANS,

MANUFACTURER OF

ARTIFICIAL LIMBS,

Deformity Apparatus, Braces,

Trusses, Etc., Etc.

89 RANDOLPH ST.,

Rooms 7 and 8.

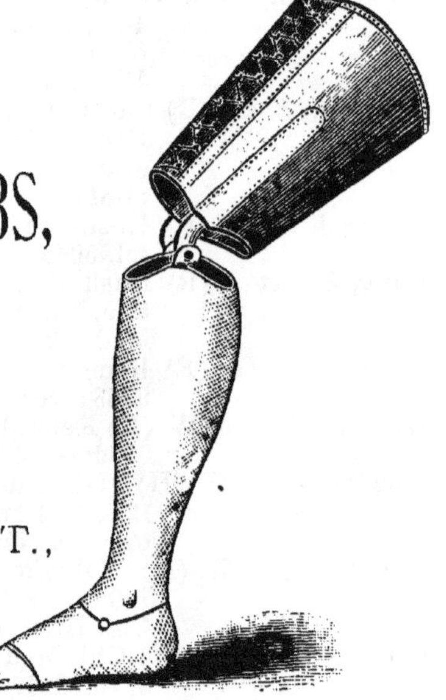

Northwestern Dental Infirmary

ROOM 210,

CHICAGO OPERA HOUSE BLOCK.

To the Medical Profession:

Your attention is invited to our institution, established some months since. None but skilled workmen are employed—no students or apprentices—and nothing but first-class work done. All extracting free, and all fillings, crown work, etc., is done at the smallest possible price. As we buy all our materials in large quantities, for cash, we are enabled to do work for what other houses are compelled to pay for the material alone.

On account of our large force of mechanical dentists, etc., we are enabled to do work in the shortest possible time, and at the least inconvenience to our patrons.

In this age of fancy prices and cheap student work at so-called dental colleges, physicians can serve the best interests of their patients by recommending them to a place where first-class work can be obtained at little more than the cost of materials, and done promptly. No waiting.

We can make artificial teeth in less time, and for less money, than any other place in the United States.

H. C. MAGNUSSON,

——————President.

Oakwood Springs

SANITARIUM,

Lake Geneva, Wis.

A View on Lake Geneva.

"LAKE ELBA," WITHIN THE GROUNDS OF OAKWOOD.

THE SANITARIUM is in the midst of a region, 20 by 30 miles in extent, having the smallest death rate of any such area in the United States. Malaria is here unknown.

Lake Geneva is 9 miles long and 200 feet deep. It has no inlet, save the pure Silurian Springs that gush from its bed and banks.

In summer, the excursions by steam, sail and row-boats; the drives, fishing and outdoor amusements in the perfect grounds of Oakwood, afford recreation. In cold weather, the Sanitarium is a perfect home of winter comforts.

For Medical Staff, class of patients treated, address, etc., please see pages 16, 121, 137, 183 and 190, in this Directory.

COOK COUNTY PHYSICIANS OUTSIDE OF CHICAGO.

BARRINGTON.

Dornbusch, H. W. (R) Rush, 1883; also at Arlington Heights, Tuesday, Thursday and Saturday.
Filkins, Frank S. (R)
Kendall, C. H. (E) Bennett, 1878; 12 to 3.
Richardson, D. H. (Rational) Rush, 1882; Chi. Hom. Med., 1883; 8 to 10 a. m., 1 to 2 p. m.
Smith, Dexter A. (R) Georgetown, 1884.

BREMEN.

Bishop, C. W. (E) Bennett, 1873; New Bremen.
Padotzckie, Mary E. (E) Bennett, 1883; New Bremen, 8 to 10, 5 to 7.

CALUMET.

Day, Frances O. (R) Wom. Med., 1884; Blue Island.
Faber, Carl (H) Chi. Hom. Med., 1879; Blue Island.
German, Wm. H. (R) Mich., 1883; Morgan Park, 8 to 9 a. m., 5 to 7:30 p. m.; 163 State, Chicago, 11:30 to 2.
Harmon, J. W. (R) Albany, 1844; Blue Island.
Heffron, Helen M. (H) Hahnemann, 1883; Washington Heights.
Kauffman, J. S. (R) Rush, 1875; Blue Island, 8 to 9 a. m., 1 to 2, and 6:30 to 7:30 p. m.
Lowenthal, Louis (H) Chi. Hom., 1879; Washington Heights.
Oliver, Emma L. (R) Wom. Med., 1881; Fernwood, morning and evening; office, Auburn, 9 to 11; tel. 88.
Oliver, W. H. (R) Victoria, Canada, 1866; Washington Heights, 9 to 12 and evenings; res. Fernwood, till 9, 4 to 6 p. m.
Pease, F. Olin (H) Chi. Hom., 1886; Morgan Park, 8 to 9, 3 to 4, and evening.

CICERO.

Bemis, J. G. (E) Bennett, 1883; P. & S.; N. Y., 1865; 300 Maple ave., Oak Park; office, 103 State, Chicago, 12 to 1. Prof. Obstetrics, Bennett Col.
Bond, Arthur G. (R) Rush, 1878; Oak Park, till 9 a. m., 1 to 2, and 7 to 9 p. m.
Hardy, Anna C. (H) Hahnemann, 1888; 310 Oak Park ave., Oak Park, 8 to 11, 6 to 7:30.
Jones, Charles E. (R) Bellevue, 1876; 118 Pine ave., Austin, 8 to 9, at 12, 7 to 8.

Lackey, R. M. (R) Rush, 1861; 115 Marion, Oak Park, 7 to 9 a. m., 1 to 2.
Latimer, H. H. (E) Bennett, 1885; Moreland.
Newell, R. C., Austin.
Pope, Ira E. (R) St. Louis, 1881; Moreland.
Stratford, H. K. (E) Eclectic, Phila., 1865; Austin, 7 to 9:30 a. m., 7 to 9 p. m.; also 243 State, Chicago, 11 to 5.
Tape, J. W. (R) Rush, 1870; 350 Lake, Oak Park, 7 to 9, 4 to 6.
Wood, E. W. (H) Geneva, 1850; 122 Marion, Oak Park, 8 to 9 a. m., 1 to 2 p. m., and evening.

EVANSTON.

Bond, Thomas S. (R) Chi. Med., 1867; P. & S., N. Y., 1868; 33 Chicago ave., till 10 a. m., 1 to 3, and at 7 p. m.; tel. 55.
Bradley, William (R) Geneva, 1864.
Bragdon, Merritt C. (H) Hahnemann, Phila., 1873; 709 Chicago ave., 8 to 9, 1 to 2, 5 to 7; tel. 36.
Brayton, Sarah H. (R) Free Med. Col. for Wom., N. Y., 1875; 607 Orrington ave.
Burbank C. H. (R) Rush, 1872; Rogers Park.
Burchmore, John H. (R) Harvard, 1875; 1005 Davis.
Clapp, E. P. (H) Hahnemann, 1882; 323 Davis.
Clayton, Allan B. (R & H) P. & S., Ont., 1869; 627 Chicago ave., 8 to 10, 1 to 2, and 6 to 8; tel. 107.
Craig, J. D. (H) Hyg. Ther. Col., N. Y., 1858; Rogers Park; also Central Music Hall, Chicago.
McCrillis, Mary F. (H) Boston Univ., 1882; 706 Chicago ave., 8 to 9, 3 to 5.
Mann, O. H. (H) 513 Davis.
Maxson, O. T. (R) Rush, 1849; 11 Ducat Block, 1 to 3; Ridge and Lincoln ave., S. Evanston.
Parker, A. H., 639 Hinman ave.
Poole, Isaac (R) Berkshire, 1862; 531 Ridge.
Romig, S. V. (R) Univ. Mich., 1872; Rogers Park.
Webster, Edward H. (R) Chi. Med., 1877; 236 Chicago ave.
Whitfield, Geo. W. (R) 316 Davis; tel. 114.

HYDE PARK.

Andrews, Sarah W., (H) Hahnemann, 1882; 1301 Bowen ave.
Arnold, Martin B., South Chicago.
Arnold, Mrs. Wilhelmina, South Chicago.
Baldwin, Olivia A. (H) Hahnemann, 1886; Univ. Mich., 1887; 4012 Cottage Grove ave.; tel. 9850.
Barrows, Ransom M. (H) Mich Univ., 1877; Hahnemann, 1884; 2702 63d, 8 to 10, 1 to 3, 6:30 to 7:30, Sunday 3 to 4; res. 6413 Sheridan ave.; tel. 9889.
Bass, Geo. E., South Chicago.
Bass, L. G., Roseland.
Benson, Fredk. N., South Chicago.

Bowerman, Mrs. Martha A. (H) Hahnemann, 1882; 1316 Oakwood boul., till 9, 1 to 4, 6 to 7:30; tel. 9850.
Boyd, R. D., Rush, 1878; 3946 Cottage Grove ave.
Brown, Thomas H., 814 43d
Buchanan, C. H. (R) Rush, 1876; 3901 Cottage Grove ave., 9 to 10, 3 to 5, 7 to 8; tel. 9849; res. 1337 40th.
Burry, Jas. (R) Chi. Med. 1875; 4012 Ellis ave.
Chapman, Geo. H. (R) Rush, 1874; 7510 Greenwood ave., till 8, 2 to 4, 6 to 7.
Coe, Milton F. (R) P. & S., 1888; 144 Oakwood boul., 10 to 12, 3 to 5; tel. 9966; res. 221 E. 42d, till 9, 12 to 1:30, after 7; tel. 9913.
Collins, R. G., (R) Chi. Med., 1885; 5103 State.
Cook, J. C. (R) Chi. Med., 1880; 130 53d, 11 to 12, 4 to 5; tel. 9829; res. 5708 Rosalie Ct., before 9, 1 to 2, 7 to 8.
Cory, A. L., (E) Bennett, 1871; 4136 Wabash ave.
Crary, Charles W. (H) Jefferson, 1858; 83 47th, 8 to 10, 2 to 4; tel. 9954.
David, J. C., 1421 Oakwood boul.; also 103 State
Dorland, E. H., 4329 Lake ave.; also Chi. Opera House bldg.
Dorn, Gay (R) Chi. Med., 1883; Cottage Grove ave. and 43d, 8 to 9, 5 to 6; tel. 9913; res. 1500 E. 41st, 12 to 2, 6 to 7; tel. 9912.
Edgar, Wm. H., 3901 Cottage Grove ave.; res. 4246 Champlain ave.
Ewing, Alice A., (H) Hahnemann, 1887; 144 Oakwood blvd; 9 to 12, 7 to 8.
French, Mrs. Amelia J., Grand Crossing.
Froom, Albert E., 3901 Wentworth; also 3726 LaSalle
Flood, J. Ramsay (R) Jefferson, 1866; res. 5320 Jefferson ave.; tel. 9829.
Frothingham, H. H. (R) Chi. Med., 1885; 4306 Lake ave., 9 to 10:30, 4:30 to 6; tel. 9982; also 235 State.
-Garceau, Alex. E., Univ. Vt., 1881; 136 53d, 9 to 10, 2 to 3, 7 to 8; tel 9829.
Gee, W. S. (H) Hahnemann, 1881; 109 53d, 8 to 10, 1 to 3:30, 6:30 to 7:30; res. 5317 Jefferson ave., till 8, after 7:30; tel. 9918.
Goetz, Julius A., Pullman.
Goodrich, A. A. (H) Hahnemann Med. Col., 1884; 4337 Champlain ave.
Graves, Kate I. (H) Hahnemann, 1885; 5730 Madison ave., 9 to 11, 6 to 8; tel. 9933.
Hall, C. B. (H) Hahnemann, 1886; 5516 Jefferson ave., 8 to 9, 6 to 7; tel. 9933.
Heissler, W. (R) Dorpot, Russia, 1883; Bacon Block, South Chicago, 8 to 10, 1 to 3, 6 to 8.
Hews, C. D. (R) Mich. Univ., 1870; Roseland.
Hibbard, William N. (R) Chi. Med. Col., 1886; 130 53d, 9 to 10, 4 to 5:30; tel. 9829; res. 5000 Greenwood ave., till 9, evening; tel. 9885.
Hilton, G. V. (R) Detroit Med. Col., 1876; 2710 63d, 11 to 12, 3 to 5; tel. 9889; res. Woodlawn Park.
Holcomb, Fred. (R) Albany, 1882; Holcomb Block, Kensington, 8 to 9, 1 to 2, 7 to 8; res. 504, 115th.

Holmes, Bayard, Chi. Hom. Med., 1884; Chi. Med., 1888; 1535 Bowen ave., till 10, after 4; also 125 State.

Huffaker, T. S. (H) Hahnemann, 1884; 3900 Cottage Grove ave., 9 to 10, 2 to 4, 7 to 8; tel. 9817; res. 245 E. 43d, till 8:45, 11 to 1, 5 to 6; tel. 9913.

Johnson, W. S. (H) Hahnemann, 1868; 5324 Washington ave., 7 to 9, 1 to 2, 7 to 8; tel. 9916.

Kemp, N. C. (H) Hahnemann, 1886; 3904 Indiana ave., 8 to 10, 1 to 3, 6:30 to 7:30; tel. 9941.

Kidder, F. H. (R) Long Island Col. Hosp., 1884; 92d and Houston, So. Chicago, 12 to 2, 6 to 8.

Lackersteen, Mark H. (R) Royal, London, 1869; St. Andrew's, 1858; 3962 Cottage Grove ave., 8 to 10, 6 to 8; tel. 9850; also, 163 State.

Larkin, J. J., Commercial blk., So. Chicago.

Lawless, James, (R) Rush, 1877; 5035 State.

Lewis, Denslow (R) Univ. Mich., 1878; cor. Lake ave. and 53d, 11 to 12, 5 to 6; tel. 9810; res. 217 53d; tel. 9844. Obstetrician and Gynæcologist, Cook Co. Hospital.

Love, John T., So. Chicago.

Lowe, Julia R. (H) Chi. Hom. Med., 1881; 3946 Lake ave., 1 to 2, 6 to 7; tel. 9856.

MacCracken, W. P. (H) Hahnemann, 1887; cor. 43d and Lake ave., till 9, 12 to 2, 6 to 7:30; tel. 9982.

McDonald, Jas. H. (R) Chi. Med., 1876; Hegewisch.

McLean, Jno. (R) Rush, 1863; 3 Florence boul., Pullman, 7 to 9, 12 to 2, 6 to 8; tel. 20.

Maull, W. C. (R) Louisville, 1869; 3979 Drexel boul., 8 to 9, 1 to 2; tel. 9817.

Meeker, Lysander (R) Keokuk, 1857; cor. Bond and 78th, Cheltenham, 6 to 8 p. m.; also 134 E. Van Buren.

McLaughlin, A. W., Houston ave., near 91st, So. Chicago.

Merrill, Arabella (H) Univ. of Mich., 1887; 4012 Cottage Grove ave., 8 to 10, 1 to 3, 6:30 to 7:30; tel. 9850.

Miller, E. (R) P. & S., 1886; 20 Kensington ave.

Morris, Alfred W. (R) Louisville Med. Col., 1877; cor. Lake ave. and 54th, 9 to 11, 1 to 2:30, 6 to 8; tel. 8835.

Nelson, H. H. (R) Rush, 1888; 4250 Cottage Grove ave., 9 to 12:30, 2 to 5; tel. 9913,

O'Neal, J. F. (E) Med. Inst., Cincinnati, 1874; 109 E. 51st, till 9, 12 to 2, after 6; tel. 9807.

Otto, Joseph P., So. Chicago.

Paine, A. G. (R) Univ. N. Y. City, 1877; 3965 Cottage Grove ave., 10 to 12, 4 to 6; tel. 9849; res. 3964 Drexel boul.

Parsons, Geo. F., (R) P. & S.. 1886; 3904 Cottage Grove ave., 8 to 10, 3 to 6, 7 to 8:30; tel. 9817; res. 3933 Drexel boul.

Pattison, J. M. Lyon (H) Hahnemann, 813 Arcade Row, Pullman, 1 to 5.

Pease, Hiram L., (R) Chi. Med., 1877; 7538 Greenwood ave., till 8, 12 to 2, 6 to 7; tel. 9821.

Peaslee, Clara W. (H) Hahnemann, 1886; 5306 Jefferson ave., 9 to 12, 3 to 5; tel. 9859,

Pusey, Chas. M. (R) Louisville, 1884 ; S. E. cor. 75th and Drexel boul., 9 to 10; tel. 9821; res. 7011 Stoney Island ave., 7 to 8:30, 12 to 1:30, after 6:30.
Randell, Geo. H. (R) Univ. Mich., 1878; 4024 Drexel boul., 8 to 9, 4 to 5; tel. 9809.
Rankin, A. C., Starling 1852; 9 Market, Pullman, till 12; 202 State, Chicago, 2 to 5.
Rockwell, C. B. (H) Chi. Hom. Med., 1883; 180 53d, Hyde Park, 8 to 10, 1 to 3, 7 to 9; tel. 9829.
Rockwell, H. O. (R) Chi. Med., 1881; 144 Oakwood boul., 10 to 12, 3 to 5, 8 to 9; tel. 9966; res. 3805 Johnson Pl.
Schmidt, F. W. (R) Chi. Med., 1885; Riverdale, till 9, 11:30 to 1:30, after 4.
Scovel, W. C., Grand Crossing.
Smith, Julia M. (H) Hahnemann, 1884; 1110 Bowen ave., till 9, 1 to 2, after 6; tel. 9912.
Sparrow, Hannah Steele (E) Bennett, 1887; Erie ave., So. Chicago, till 9, after 12 m.
Stebbings, Horace P. (R) Chi. Med., 1886; 130 53d, 1 to 3; 6110 State, 9 to 10, 5 to 6, 8 to 9; tel. 9829; res., 6015 Indiana ave., till 9, 12 to 1, and nights. Health Officer, Hyde Park.
Stockwell, J. S. (R) Rush, 1888; 144 Oakwood boul., tel. 9850.
Stone, Willis C. (R) Rush, 1884; 4258 Cottage Grove ave., 9 to 11; tel. 9913; res. 4005 Prairie ave., afternoons; tel. 9941; at Central Free Dispensary, 2 to 4 Tuesdays and Fridays.
Swan, Chas. F. (R) Ohio Med., 1875; So. Chicago, till 9, 12 to 2, 6 to 8; tel. 8941.
Tillotson, H. Jno. (R) P. & S., Baltimore, 1880; Hegewisch ave. and 133d, Hegewisch, 7 to 9 a. m., 7 to 9 p. m.
Wagner, Mrs. Caroline, So. Chicago.
Waite, Lucy (H) Hahnemann, 1883; cor. 53d and Jefferson ave., 10 to 2; tel. 9859.
Wardner, M. S. (R) Rush, 1884; 4201 Cottage Grove ave.
Was, J. W. (R) Univ. Mich.; Roseland, 7 to 9, 2 to 4, 6 to 8.
White, G. J. (R) Toronto Med. Col., 1885; 5349 Lake ave., till 9:30, 2 to 3.
Webster, E. M. (R) Starling, 1884; Winnipeg blk., So. Chicago, till 10, 1 to 3, 6 to 8.
Wetherell, G. T. (R) Univ. N. Y., 1856; 164 Oakwood boul., 8 to 9, 5 to 8; 113 Adams, 10 to 3.

JEFFERSON.

Adams, A. L. (E) Bennett, 1886; Irving and Douglass aves., Irving Park, 10 to 12, 4 to 6.
Butler, Alvin S. (H) Hahnemann, 1884; Maplewood, 10 to 12, 5 to 7.
Dutd, W. A. (E) Bennett, 1885; Maplewood.
Moore, Malcolm T. (R) P. & S., 1885; Maynard st., 8 to 9 a. m., 6 to 8 p. m.

LAKE.

Arnold, W. J. (R) Michigan, 1884; 68th and Wallace.

Bacon, E. Z. (H) Hahnemann, 1886; 5629 Wentworth ave., till 9, 3 to 6.

Bacon, Martin W. (R) Univ. of Mich., 1875; 6700 Perry ave. till 9, 2 to 4, 7 to 8.

Bell, Geo. (R) Victoria, Can., 1888; 4222 Ashland ave., 3 to 5; res. 1410 35th.

Bell, James (R) Victoria, Can., 1887; 4944 Ashland ave., 8 to 10, 12 to 2:30, 6 to 8; tel. 9543.

Blaine, J. E. (R) Bellevue, 1872; 309 63d st.

Borter, F. X. (R) Innsbruck, 1878; 4822 Ashland ave., 9:30 to 12; tel. 9543; also N. W. cor. Polk and Clark, Chicago.

Bronson, Henry (R) Univ. Vt., 1887; 6848 Wentworth ave.

Brooke, F. C. Normal Park.

Brown, Thos. H. (H) Chi. Hom. Med., 1884; 814 43d, 8 to 10, 3 to 5; tel. 9550.

Burke, R. H. (R) Univ. of Mich., 1880; 4209 S. Halsted, 2 to 4 p. m.; tel. 9601: res. 4700 Ashland ave., 7 to 9, 6 to 8.

Bushee, G. B. (H) Hahnemann, 1887; cor. Wright and Chestnut, till 9, 12 to 2, 6 to 7; tel. 86.

Caldwell, Chas. P. (R) Chi. Med., 1876; Rush, 1877; 749 W. 43d, till 9, 12 to 1, after 5.

Caldwell, H. J. (R) Trinity, 1885; 4902 State, till 9, 11 to 2, after 7; tel. 9807

Calvert, G. S. (R) Northwestern Med. Col., St. Joseph, Mo., 1885; 6913 Sherman, Englewood.

Champlin, A. H. Mich. Univ., 1869; 61st and School.

Chavett, Franklin (E) Bennett, 1869; 6333 Yale.

Clendenning, J. W. (R) 4702 State, till 9, after 7; tel. 9807.

Collins, R. G. (R) Chi. Med., 1885; 5103 State, till 9, 1 to 2, 7 to 8; tel. 9807.

Cory, Alphonso L. (R) Bennett, 1871; 4136 Wabash, 8 to 9, 1 to 3; tel. 9950.

Curry, C. P. (R) 6034 Halsted, 7 to 10, 7 to 9.

Davis, J. G. (R) 544 63d.

DeWolf, James E. (R) Harvard, 1866; 440 Englewood ave.; tel. 49.

Dougherty, Patrick (R) Chi. Med., 1884; 5101 Wentworth ave., till 9, 12 to 1, 6 to 8; tel. 65.

Dow, M. C. 4201 Wentworth ave.

Doyle, Jeremiah, Dublin 1871; 1239 47th.

Eskridge, J. H. (R) 815 43d, 8 to 10, at 12, after 6; tel 9536.

Fairbanks, C. D. (H) Hahnemann, 1867; 6342 Yale, 8 to 9, 12 to 2, 6 to 8; tel. 14.

Fenn, Thos. L. (R) P. & S., N. Y., 1881; 4631 Wentworth ave., till 9, 1 to 3, 6 to 8; tel, 9647.

Foster, J. M. (R) Harvard, 1866; Vehmeyer Block.

Foulks, C. Allison (R) Chi. Med., 1885; 4700 State, 10 to 12, 6 to 8; tel. 9807; also 279 State, Chicago.

Goodall, W. W. (R) 4626 Wentworth ave.

Goodhue, Mrs. H. A. (H) 558 62d st.
Goodhue, O. A. (H) 558 62d st.
Greenleaf, Geo. T. (H) Chi. Hom. Med., 1881; 6557 Wentworth ave., 8 to 9, 12 to 3, 6 to 8; tel. 39.
Hall, J. L. S. (R) Bennett; 4704 Ashland ave., 10 to 12, 2 to 4.
Harris, A. F. (H) Hahnemann, 1884; 514 61st till 9, 12 to 2, 6 to 7.
Heskett, S. F. Normal Park.
Holman, E. E. (H) Hahnemann, 1878; Behmeyer Block.
Hunt, J. S. (R) Rush, 1884; 558 63d st.
Hurford, W. D. (R) Jefferson, 1878; 6714 Lafayette ave., till 9, after 4:30.
Hurst, N. N. (R) Jefferson, 1873; 3906 State, 8 to 10 a. m., 1 to 2, 8 to 9 p. m.; tel. 9853; at 5100 Wentworth ave., 4 to 5.
Jones, Joshua (R) Mich. Univ., 1879; Auburn, till 8, 12 to 2, after 8.
Kirkpatrick, Lafayette, Ohio Med., 1879; 6246 Wentworth ave., 7 to 9, 2 to 4, after 7.
Lawless, James, (R) Rush, 1877; 5035 State.
Lovewell, C. H. (R) Univ. of Mich., 1871; 6058 Wentworth ave, till 8, 12 to 2, after 7; tel. 6.
Lyman, M. J. (R) Castleton, 1855; 6401 Stewart ave., 7 to 9, 1 to 3
Marley, J. W. (E) 3904 State, 10 to 12, 7 to 9; tel. 9853; res. 6358 Honore.
McNeal, John, 6700 Stewart.
Liddell, Mark J., S. Englewood.
McKenna, M. J., 4062 State, 4306 Wentworth ave.
Meehan, Martin G. (R) Rush, 1885; 4341 S. Halsted, 8 to 10, 12 to 2, 6 to 8; tel. 9550.
Minaker, W. (R) Royal Edin.; 6204 Wentworth ave., 8 to 10 2 to 4, 7 to 9; tel. 58; res. 6206 Wentworth ave.
Morton, E. C. (R) Chi. Med., 1888; 6801 S. Halsted, 8 to 9, 1 to 2:30 7 to 8:30; tel. (Englewood) 66.
North, Charles F. (R) Univ. of Leipsic, 1884; 309 63d, 9 to 10, 2 to 3, 7 to 8; res. 240 61st, 1 to 2.
Oliver, Emma L. (R) Wom. Med., 1884; Auburn 9 to 11; res. Fernwood.
Parsons, Wm., 4335 Emerald ave.
Pierpoint, Ernest, 6553 Yale.
Poppe, Otto (H) Hahnemann, 1870; 4423 Wentworth av.
Reasner, Marie E. (E) Bennett's 1878; 5729 LaSalle, till 9, 5 to 6; also Cent. Music Hall, 11 to 2.
Reese, Joseph (R) 5108 and 5420 S. Halsted.
Rowe, Adeline A. 6223 School.
Saguin, Hubert cor. 58th and Wright; tel. 5750.
Spach, A. B. (H) Chi. Hom. Med., 1886; 6310 Wentworth av.
Sparling, Ellis H. (H) Chi. Hom. Med. 1884; 6737 Yale.
Stockham, Alice B. (H) 6058 LaSalle, also 161 LaSalle st., Chicago.
Strickland, Chas. O. (R) Rush, 1888; 1043 53d till 9, 11 to 12, 2 to 3; Branch office 5108 S. Halsted, 9 to 10, 4 to 5.
Sullivan, T. J. (R) Univ. Mich, 1880; 4209 Halsted, 2 to 4; res. 4700 Ashland av.; tel. 9601.
Sweet, H. P. cor. School and Chestnut.

Tallman, E. D. (R) Univ. of Mich., 1882; Auburn Park, 2 to 4, tel. 88; res. S. Englewood till 9, 1 to 2, after 7; tel. 87.

Taylor, Cora E. (H) Pulte, Cincinnati, 1884; 6356 Stewart av.

Thornton, Francis E. (E) Bennett, 1888; cor. Belmont and Ellis avs. 7 to 9, 5 to 8; res. Avondale.

Tillotson, G. K. (A) Rush, 1876; 4132 Wentworth av. till 9, 12 to 2, after 7.

Turner, B. S. (R) Chi. Med., 87; 3906 State, 12 to 1, 3 to 5, 8 to 9; tel. 9853; 5100 Wentworth av., 10 to 11; tel. 35, (Englewood.)

Westerfield, W. C. (R) St. Louis, 1859; 553 Englewood av.

Whitford, Henry Edgar, (E) Bennett, 1881; Newman blk. 63d and Stewart av. till 9:30, 12 to 1:30, after 6 tel. 33 (Englewood). Prof. of Dermatology and venereal Diseases, Bennett Med. Col.

Whitman, Chas. H. 5765 Wentworth av.; also 119 Madison, Chicago.

Wilder, D. J. (R) Univ. of Mich., 1877; 6320 Wentworth av. till 8:30, 1 to 3, 6 to 8.

Wolford, Wm. H. 3904 State.

LAKE VIEW.

Abbott, W. C. (R) Univ. Mich., 1885; Ravenswood till 9, 1 to 2, after 6 tel. 12106; Argyle 3 to 4.

Avery, Elizabeth (H) Hahnemann, 1887; 315 Lincoln av. 8 to 10:30, 6:30 to 7:30.

Bacon, J. V. (R) Chi. Med., 1885; Lincoln and Belmont avs. 4 to 5 P. M.; tel. 12075; res. 1402 Wrightwood av. 1 to 2, 6 to 7, and nights; telephone 12035.

Barker, W. A. (H) Hahnemann, 1879; 1740 Diversey av. 8 to 10, 2 to 4, 7 to 8; tel. 12033.

Barnet, A. D. (R) Toronto, 1887; 721 Lincoln av. till 9, 1 to 2, after 7; tel. 12064.

Bennett, E. R. (R) Rush, 1882; 893 Clybourn av. 7 to 8 a. m., 3 to 4 p. m.; tel. 12,038.

Blake, S. C. (R) Harvard, 1853; 576 Fullerton av. 4 to 6 p. m.; tel. 3966, also 125 State, Chicago. Consulting Staff Women and Children's Hospital.

Bridgeford, Jennie E. (R) Wom. Med., 1877; 1426 N. Clark till 10, after 5, tel. 12033. Physician to Floating Hospitals.

Brigham, L. Ward (E) Bennett, 1888; Gross Park, 10 to 12, 2 to 4, 6:30 to 7:30; res. 754 Herndon; tel. 12075.

Bunn, M. O. (R) 1619 Diversey av.

Cross, Edwin (H) Cincinnati Col. M. & S., 1872; 1436 Wrightwood av., 8 to 9:30, 1 to 2:30. 6 to 7:30; tel. 12087.

Gates, W. S. (R) Chic. Med. 1881; 1247 Wrightwood av. till 9, 12 to 1, 6 to 7; tel. 12020.

Goldsborough, C. B. (R) Univ. of Penn., 1876, Resident Surg. in Charge U. S. Hosp., tel. 3907.

Goodsmith, H. M. (R) P. and S., 1887; 413 Lincoln av. till 10, 12 to 2; 12041.

Goodsmith, Wm. P. (R) Rush, 1883; 413 Lincoln av.

Green, Isadore (H) Hahnemann, 1886; 315 Lincoln av. 9 to 12, 7 to 8; tel. 3317.

Helmuth, C. A. (R) Berlin, 1844; 624 Lincoln av., also 26 N. Clark.

Hobart, W. F. (H) Chi. Hom. Med., 1886; Wood and Melrose, Gross Park, 8 to 9, 1 to 2, 6 to 8; tel. 12039.

Hoffman G. A. (E) Bennett, 1878; 838 Seminary ave.

Jacobs, J. M. (E) Bennett, 1887; cor. Lincoln and Southport avs. till 9, 1 to 2, after 5; tel. 12124.

Keeton, Theodore A. (R) Jefferson, 1876; Ravenswood 10 to 2; tel. 3908.

Ludwig, C. H. (H) Chi. Hom. Med., 1875; 600 Lincoln av. till 9, after 3; tel. 12035.

McKittrick, Elizabeth (R) Wom. Med., 1880; 450 Racine av., 8 to 10, 4 to 6; tel 3959.

Oliver, N. E. (R) Chi. Med., 1880; 1373 N. Clark.

Palmer. A. E. (R) Rush, 1885; 721 Lincoln av., Lake View, till 8:30, 12 to 2, night; tel. 12064.

Parsons, H. E. (E) E Ravenswood Park.

Pingree, M. G. (E) Bennett, 1880; Ravenswood, also 103 State, Chicago.

Porter, F. D. (R) Detroit, 1877; 1594 N. Halsted, till 9, 12 to 1, 5 to 7; tel. 12118.

Printy, J. A. (H) Univ. Iowa, 1882; 598 Lincoln av.. till 9, 12:30 to 2, after 6; tel. 12035.

Pursell, P. H. (R) Univ., Penn., 1864; Ravenswood, 7 to 10, 1 to 3, 7 to 9; tel. 12108.

Reid, T. J., Ravenswood; also Opera House bldg., Chicago.

Rowe, W. C. (H) Hahnemann, 1883; 1305 Wrightwood ave., till 9:30, 1 to 2:30; 6 to 7:30; tel. 12035.

Rutherford, C., 102 Fullerton ave.

Scholer, E. C. (H) Chi. Hom. Med., 1888; 886 Lincoln ave., 10 to 12, 2 to 4, 5 to 7; tel. 12039.

Scholar, J. H. (R) Berne, 1850; 886 Lincoln ave.; till 11, after 3; tel. 12039.

Semple, W. F. (R) Rush, 1881; 1301 Belmont ave., 7 to 9, 1 to 2, after 7; tel. 12037.

Sieber, F. A. P. (R) Rush, 1883; 429 Lincoln ave., till 9, 12 to 2, after 7; tel. 12012. City Physician, Lake View.

Smith, Geo. S. (R) Jefferson, 1856; 1202 Diversey ave., 8 to 9, 1 to 3, 7 to 8; tel. 12045; res. 1722 Diversey ave., till 8, after 9.

Walker, W. S. (R) Rush, 1886; 722 Lincoln ave., 11 to 3, after 6; tel. 12030.

Weil, Carl A., 1107 Lincoln ave.

Whitnall, W. R. (R) P. & S., 1884; 1066 Lincoln ave., 9 to 11, 2 to 4; tel. 12075.

LEMONT.

Cook, William B. (E) Bennett, 1882.

Fitz Patrick, J. A. (R)

Leahy, John (R) Rush, 1885.

Roberts, S. A. (H) Hahnemann, 1877.
Thorpe, J. C. (R)
Von Bernauer, J.

LYONS.

Atwater, John (H) Hahnemann; Western Springs.
Carey, Warren (R) Rush, 1883; La Grange.
Congdon, J. L. (R) Rush, 1865; office, Lyons; res. Riverside.
Fox, Geo. M. (R) Castleton, 1851; La Grange.
Higgins, Arthur E. (R) Rush, 1886; La Grange.
Jerome, Levi R. (R) La Grange.

MAINE.

Carrier, Chas. W. (H) Chi. Hom., 1877; Desplaines.
Fricke, Gustave H. (R) Rush, 1869; Park Ridge.
Geltch, Ernest A. (R) Rush, 1888; Desplaines.
Hammond, John H. (H) Hahnemann, 1887; Park Ridge.
May, Jacob (R) Rush, 78; Desplaines.
Thielo, Henry C. (R) Rush, 1884; Desplaines.

NEW TRIER.

Morrison, Geo. H. (H) Hahnemann, 1881; Winnetka.
Nelson, Jas. W. (R) Chi. Med., 1886; Winnetka, 11 to 12, 5 to 6.
Stolp, Byron C. (E) Bennett, 1873; Wilmette.

NILES.

Hoffman, G. F. T. (R) Rush, 1861.

NORTHFIELD.

McCornack, E. A. (E) Bennett, 1881; Oak Glen, 7 to 9, 5 to 6; at Shermer, Monday and Thursday, 9:30 to 11.

NORWOOD PARK.

Hughes, J. O. (R) Rush, 1867.

ORLAND.

Schussler, W. G. (E) Bennett, 1885.

PALATINE.

Hulett, S. E. (H) Hahnemann, 1876; Plum Grove ave. and Slade, 7 to 9 a. m., 1 to 2 p. m.
Pearman, Jas. O. (R) Rush, 1885.
Wadhams, (R) Rush, 1878.

PROVISO.

Clendennen, Irving (E) Bennett, 1872; Maywood.
Clark, F. D., Maywood.
Coryell, George (R) Riverside.
Hotchkiss, Mrs. Isabella S. (H) Chi. Hom. Med., 1880; Maywood.
Kahle, Franz T. (H) Hahnemann, 1868; Harlem.
Munger, M. Jerome (R) Georgetown, 1865; Maywood.
Roberts, Walter C. (R) Rush, 1881; 14 N. 5th ave., Maywood, 8 to 10, 2 to 3, 7 to 8.

THORNTON.

Doepp, Wm. (R) Giessen, 1856; Homewood.
Oliver, Nelson E. (R) Rush 1880; 7 to 9 a. m., 12 to 2 p. m.
Weidner, M. Robert (R) P. & S., 1883; Dolton.

WAYNE.

Guild, W. L. (E) Bennett, 1884; at office, 8 to 10, 7 to 9; at res. 10 to 2.

WHEELING.

Best, John E. (R) Rush, 1870; Arlington Heights, 8 to 9 a. m., 12 to 1 p. m.
Dornbusch, H. W. (R) Rush, 1883; Tuesday, Thursday and Saturday at Arlington Heights; res. Barrington.
Hawkes, J. B. (R) Berkshire, 1848; Arlington Heights.

WORTH.

(For Blue Island physicians see Calumet.)

Department of Dental and Oral Surgery of Lake Forest University.

Northwestern College of Dental Surgery

Southeast Cor. Wabash Ave. and Twelfth St.,

CHICAGO, ILLINOIS.

WM. C. ROBERTS, D.D., LL.D., PRESIDENT.

F. H. B. McDOWELL, ACTUARY.

FACULTY.

G. C. PAOLI, A.M., M.D., Emeritus Professor of Materia Medica.

N. P. PEARSON, A.M., M.D., Emeritus Professor of Pathology.

ENOS J. PERRY, D.D.S., Professor of Operative Dentistry.

JOSEPH HAVEN, M.D., Professor of Physiology and Diseases of the Nervous System. Dean of Faculty.

BYRON D. PALMER, D D.S., Professor of Prosthetic Dentistry.

J. H. LYON, A.M., M.D., Professor of General and Dental Pathology.

J. H. SALISBURY, A.M., M D., Professor of Chemistry.

J. E. HEQUEMBOURG, M.D., Professor of Anatomy and Principles and Practice of Surgery.

F. C. CALDWELL, M.D., Professor of Materia Medica and Therapeutics.

NORMAN J. ROBERTS, D.D.S., Professor of Oral Surgery.

This College offers to students of dentistry a course of instruction ranking with the best dental schools in the country, with the addition of one of the largest clinical practices in the United States. Over eleven thousand different operations were performed in its operating rooms during the past year, giving its students unusual facilities for the practice necessary to a complete education in this specialty of medicine.

The fourth annual winter session opens on the first Wednesday in October and continues six months. Graduates in medicine may have its degree conferred after one year's practical experience in dentistry and attendance upon one winter's course of lectures.

ANNUAL FEES, $100.

For other information and Catalogue, address

F. H. B. McDOWELL, ACTUARY,

1201 Wabash Avenue, Chicago, Illinois.

OAKWOOD SPRINGS SANITARIUM,

LAKE GENEVA, WIS.

A FOREST DRIVE

Within the grounds of OAKWOOD SPRINGS SANITARIUM.

WHILE this park lies wholly within the city limits, it is by nature so bounded as to render the entire park absolutely private, affording patients the most perfect seclusion and at the same time the utmost freedom of action and enjoyment. It has a third of a mile frontage on Main street, and extends northward more than a quarter of a mile. It lies also within five minutes' walk of the post-office, railway station and hotels. Across one corner of the park flows the swift and sparkling creek, or *White River*, the outlet of Lake Geneva, which, in its course, forms Lake "Elba" within the park (see page 170), twenty feet in its deepest soundings, and covering an area of twenty acres. It is confidently believed that there is not another site, in the whole country, equaling in beauty and so perfectly adapted to the requirements and convenience of such an institution.

Please see pages 16, 121, 137, 170 and 190.

CHICAGO DENTISTS.

Abbott, Lawrence F., Ivar Flats, suite B.

Abbott, Mrs. M. E., Ivar Flats, suite B.

Allport, Walter W., (M. D. Rush) 7 Jackson, room 23, 10 to 4; tel. 2570; res. 69 Maple. Dental Surgeon, St. Luke's Hosp.

Ames, Wm. B., Ohio Col. Dent., Surg., 1880; 70 State, 9 to 5; res. 917 Pullman bdg.

Aspinwall, M. H., 10 Central Music Hall.

B

Bacon, Dewitt C., 281 Lincoln ave.

Baker, Benj. M., 103 State, room 28.

Baker, Chas. R., 3 Belvedere building.

Baldwin, A. E., Chi. Den. Col. '84 (M. D. Rush) 828 W. Adams.

Ballard, Chas. W., 800 S. Halsted.

Barclay, J. C., 355 Dearborn ave.

Barnes, Charles T., 13 Central Music Hall.

Barnum, Henry Leon, (H) Chi. Homeo. Med. Col., 1883; Chi. Col. Dent. Surg., 1885; 628 W. Lake, 8 to 3; tel. 7016.

Becker, George H., Chi. Dent. Col. 1888; S. E. cor. Fullerton and Lincoln aves., 9 to 5.

Benham, Emma L., 1 Park row.

Bentley, Geo. H., 70 Dearborn, room 7.

Bently, Chas. E., 279 State.

Bentzen, Michael H. 1002 W. Madison.

Bergman, Gustaf, 93 Sedgwick.

Bills, Geo. H., 70 Madison, room 5.

Bischoff, Christian, 174 North ave.

Blackburn, Robert, 4251 S. Halsted.

Bohan, John C., Blue Island ave. and 12th.

Boulter, H. H. 70 State.

Bowman, Frederick H., 1412 N Clark.

Brass. O. H., 236 State.

Broadbent, Thos. A., 858 Clybourn ave.

Brophy, Truman W., (M. D. Rush) 96 State, room 501. Professor Dental Pathology, Rush; Oral Surgery, Chi. Col. D. S.

Brown, A. E., 1536 Wabash ave.

Brown, Gilbert T., 73 Clark.

Buell, Harry C., 3900 Cottage Grove ave.

Burlingham, James S. 238 N. Clark.

Bush, Lorenzo, 125 State, room 13.

C

Cady, Edward E., Chi. Col. Dent. Surg., 1885; 240 Wabash ave., 9 to 5; tel. 2457; res. Leland Hotel.

Cady, G. B. 163 State, room 40-41.

Cain, W. H. H. 112 Dearborn.

Calkins, Chas. D., 225 Dearborn, room 706.

Carson, Clayton W., cor. Cottage Grove and 39th.

Cattell, David M., 271 Wabash ave.

Celley, F. M., 291 31st.
Christmann, Geo. A., 95 5th ave.
Cigrand, B. J., 1056 Milwaukee ave.
Cigrand, Peter J., 290 W. 12th.
Clapp, James L., 183 Clark, room 4.
Clark, Albert B., 89 Madison, room 10.
Cleveland, Chas. E., 241 Wabash ave.
Cleveland, M. B., 126 State, room 25.
Clifford, E. L., Penn. Col. Dent. Surg., 1879; 454 W. Madison, 9 to 5; tel. 4090.
Clusmann, L., 360 Blue Island ave.
Costner, Henry A., 171 22nd, rooms 3, 4 and 5.
Craig, Joseph, 374½ N. Market.
Crissman, Ira B., 271 N. Clark.
Crouse, J. N., 2231 Prairie ave.
Cummins, Wm. G., 70 State, room 300.
Curtis, Wm. W., 235 State.
Cushing, Geo. H., 96 State, room 514.

D

David, V. R., 240 Wabash ave.
Davis, E. E., 524 W. Van Buren.
Davis, Frank H., 52 31st.
Davis, Lyndall L., 524 W. Van Buren.
Day, Wm. E., 137-143 State, 8 to 6; tel. 5604.
DeCamp, A. L., S. W. cor. State and Randolph, room 201.
Deschaner, Frederick, 95 5th ave.
Deschaner, Joseph, 95 5th ave.
Devlin, J. B,, 127 22nd.
Dillon, R. A. C., 659 Sedgwick.
Dittman, Julius W., 113 Madison.
Dryer, C. W., 471 Milwaukee ave
Dunn, J. Austin, Chi. Col. Dent. Surg., 1885; 70 Dearborn, 9 to 5; res. 329 37th. Prof. Anatomy Chi. Col. Dent. Surg.
Dyer, Arthur E., 2512 Wabash.
Dyer, Walter C., 2512 Wabash.

E

Eaton, Lewis, 235 State.
Ellis, J. Ward, 225 Dearborn, room 16.
Emmart, Chas. M., 3858 State.
Etzler, B. M., 68 35th.

F

Fahnestock, Albert, 1802 State.
Fernandez, E. M. S., 103 State, room 33.
Fowler, W. F., 812 Opera House bldg.
Freeman, Andrew W., 920 Opera House bldg.
Freeman, Arthur B., 325 W. Madison.
Freeman, I. A., 126 State, room 36.
Freund, Fred G., 150 North ave.

G

Gale, Willis H., 129 22d.
Gardiner, Frank H., 126 State, room 50.
Gardiner, Thos. D., 65 Randolph.
Goetz, Albert, 952 Milwaukee ave.
Goodman, A. G., 202 State.
Graves, Erastus L., 496 W. Madison.
Greenwood, E. N., 247 W. Madison.
Guffin, Edwin L., 125 State, room 28.

H

Hagist, Geo. M., 27 N. Clark, room 12.
Hale, Leon T., 1014 W. Lake.
Harlan, A. W., 70 Dearborn, room 9.
Harris, Andrew J., 279 Warren ave.
Hartt, C. F., 490 W. Madison; tel. 7022.

Haskell, L. P., 34 Monroe, room 32.
Haskins, Geo. W., Chi. Col. Dent., 1887; 70 Dearborn, room 10, 9 to 5; res. 524 W. Congress.
Hasselrus, Rudolph, Chi. Col. Dent. Surg., 1885; 409-411 Milwaukee ave.
Hebert, Alfred W., 96 State.
Hemmingway, Hannaniah W., 69 Dearborn.
Henkel, Albert F., 657 W. 12th.
Herrmann, Richard, Chi. Dent. Col., 1888; N. E. cor. Fullerton and Lincoln aves., 9 to 5.
Hewett, Austin, C., 491 W. Adams.
Hewitt, N. T., 999 W. Madison.
Higgins, I. E., Chi. Dent. Col.
Hindberg, Charles G., 166 Sedgwick.
Honsinger, Emanuel, 318 Park ave.
Hoyt, Alfred W., 243 State, room 73.
Hunt, Edward W., 207 Clark.
Huxman, Ernest A., 105 N. Clark
Huxman, F. W., 167 N. Clark.

I

Ireland Lewis E., 202 State.
Irey, James C., 204 W. Madison.

J

Johnson, C. N., Royal Col. Dent. Surg., Ontario, 1881; Chi. Col. Dent. Surg., 1885; 612 Chicago Opera House, 9 to 5; tel. 1294; res. 841 N. Clark. Lecturer on Operative Dent. and Pathology, Chi. Col. Dent. Surg.
Johnson, M. B., 89 Madison.
Johnson, Wm. J, 300 31st.

K

Keefe, James E., 912 Opera House bldg.
Kennicott, Jonathan A., 803 Opera House bldg.
Kester, P. J., 567 W. Madison.
Klein, Carl, 395 N. Wells.
Knapp, A. S., 164 Dearborn.
Koch, Chas. R. E. 3011 Indiana ave.
Krueger, J. F., 203 Blue Island ave.
Kuester, Wm., 11 Mohawk.

L

Lane, Asa H., 305 Division.
Lattan, Louis F., 973 Van Buren, 9 to 5.
Lauer, L. A 70 State, room 203.
Lawrence, E. E., 785 W. Madison.
Lawrence, Geo. W., 125 State, room 15.
Lawrence, Mrs. H. E., 96 State.
Lawrence, Pliny I., 78 Monroe, room 31.
Lazear, W. W., 2208 Wabash ave.
Lechner, Lester F., 329 Aberdeen.
Leggett, John, 207 Clark, room 6.
Leggo, Basil A., 188 W. Van Buren.
Lewis, Chas. W., Chi, Col. D. S., 1885; 445 N. Clark; tel. 3148.
Lewis, Geo. G., 294 Hermitage ave.
Lewis, W. F., 273 W. Madison.
Lichtenburgh, C. W., 290 W. 12th.
Lindos, J., 247 W. Indiana.
Long, Geo. E, 610 W. Adams.
Low, James E., 164 Dearborn.
Ludwig, R. F. 34 Monroe, room 32.
Lund, Chas. D., 206 Milwaukee ave.
Lutwyche, F. H., 277 S. Halsted, 8 to 8; res. 22 Blue Island ave.

M

Magnusson, Howard C., 210 Opera House bldg.
Mann, Anthony, 169 S. Clark.

Mann, Julia C., 70 Dearborn, room 8.

Marcoux, H. F., Standard Theater Bldg.

Marsh, John S., 241 Wabash ave.

Marshall, Fred C., 70 Dearborn, room 12.

Marshall, John S., M. D., Syracuse; 7 Jackson, room 23, 9 to 4; tel. 2570; res. 3343 Prairie ave. Lecturer Oral Surgery, Chi. Med. Col.; Dean and Prof. Oral Surgery, University Dental Col.; Dent. Surg., Mercy and St. Luke's Hosps.

Martin, Joseph H., 169 Clark, room 10.

Martin, Wm. J., 181 W. Madison.

Matteson, Arthur E., 3700 Cottage Grove, 9 to 12, 1:30 to 5:40; tel. 9874; res. 3822 Langley ave. Prof. Dental Irregularities and Prosthetic Dentistry, Dent. Dept. N. W. Univ.

Matteson, Chas. F., 3501 Cottage Grove.

Matthews, J. H., 245 W. Madison

McChesney, A. C. M., 73 Clark.

McChesney, James B., 73 Clark.

McChesney, Wm. B., 73 Clark.

McConnell, Whiteford, 136 Clark.

Merkley, Whitney C., 1115 W. Van Buren.

Merriman, Chas. J., 820 W. Monroe.

Moelmann, Ernst O., 858 Milwaukee.

Muser, L. J., 213 W. 12th.

N

Nelson, Arthur, Chi. Col. Dent. Surg., 1887; 70 State, 9 to 4:30; res. Desplaines.

Nelson, Geo. A., 480 Milwaukee ave.

Nelson, N., 230 Milwaukee ave.

Newkirk, Garrett, 1558 Wabash.

Newman, James L., 163 State, room 43.

Nichols, Amos J., Chi. Dent. Col., 1885; 565 W. Madison.

Nichols, Gorton W., 308 W. Adams.

Norris, Robert A., 126 Dearborn, room 2.

Norton, J. W., 96 State, room 512.

Norton, M. Eugene, 126 State.

Noyes, Edmund, Chi. Col. Dent. Surg., 1884; 65 E. Randolph; res. 24 Lincoln ave. Prof. Operative Dentistry, Chi. Col. Dent. Surg.

O

Ottofy, Louis, Western Col. Dent. Surg., St. Louis, 1879; 1228 Milwaukee ave., 8 to 12:30, 1 to 4:30; tel. 4203. Lecturer on Anatomy and Physiology, Chi. Col. Dent. Surg.; Clinic Operator, Chi. Col. Dent. Surg.

Oviatt, Albert E., 198 Honore.

P

Page, J. C., 473 Ogden ave.

Paine, R. M., 739 Larrabee.

Palmer, Byron D., 103 State, room 27.

Palmer, Byron, S., 224 Warren ave.

Palmer, David G., 372 W. Taylor

Palmer, Philip A., 125 State room 18.

Parkinson, Edwin J., Baltimore Col. Dent. Surg., 1879; 2842 State, 8:30 to 5; tel. 8185.

Peck, A. H., 122 Wabash ave.

Perry, E. J., 82 W. Madison.

Pfennig, Ernst, 18 Clybourn ave.

Pitt, H. N., 491 W. Adams.

Plattenburg, Cyrus B., 737 W. Madison.

Poessel, E., 76 McVicker Theatre bldg.; 416 W. Chicago ave.

Pruyn, Chas. P. (R) Chi. Dent. Col., 1885; Rush, 1886; 70 Dearborn, 9 to 5; res. 219 Wisconsin, Oak Park. Prof. of Principles and Practice of Operative Dentistry, Den. Col. N. W. Univ.

R

Reid, James G., 69 Dearborn, room 30.

Rhein, Benj. L., 3012 Cottage Grove.

Rice, L. C., 35 Central Music Hall.

Rice, Ogilvie A., 306 W. Indiana.

Rimmer, Rueben L., 229 W. Madison.

Robertson, John D., 718 W. Van Buren.

Rogers, James W., 818 Opera House bldg.

Rogers, J., 459 W. Randolph.

Rosenthal, William. M., 2873 Archer ave.

Rowley, Clarke R., Am. Col. Dent. Surg., 1888; 163 State, 9 to 5; res. 4225 Michigan ave. Clinical Instructor, Am. Col. Dent. Surg.

Royce, Edward A., 643 W. Monroe.

Ryan Edward P., N. W. Dent. Col., 1886; 226 Wells, 9 to 12, 1 to 5; tel. 3274.

S

Salomon, Godfrey S., 15 Central Music Hall.

Satterlee, Frank W., 199 Clark.

Schmalstig, Geo. F., 76 Clybourn ave.

Schnell, Wm., 189 Chicago ave.; 413 N. Wells.

Schoen, John, 69 Dearborn, room 5.

Schycker, M., 188 Clark.

Seeglitz, Otto E., Chi. Col. Dent. Surg., 1887; 445 North ave., cor. Wells; tel. 3206.

Sherwood, Geo. A., 3017 Wabash ave.

Smith, C. Stoddard, Philadelphia Dent. Col., 1865, 103 State, 10 to 4.

Smith, David J., 1802 State.

Smith, Geo. F., 240 Wabash, room 25.

Smith, John K., 125 22nd.

Smith, Joshua, 165 22nd. and 2459 Prairie ave.

Smith, Louis B., 268 31st.

Smith, Marvin E., 163 State, room 67.

Smith, Willoughby B., 126 State, room 48.

Sovereign, Chas. W., 108 Washington.

Sperling, Isaac D., 170 State, room 14.

Spirkel, John, 287 W. 12th.

Stansbury, Joseph A., 1860; 343 W. Monroe, 9 to 12, 1 to 4; tel. 4062.

Starr, Robert W., 64 E. Chicago ave.; 3661 Indiana ave.

Steele, Robert, 514 Opera House bdg.

Stevens, Walter A., 2631 Wabash.

Stewart, Chas. F., 219 31st.

Stewart, Henry, 9 31st.

Stewart, James, 3035 S. Park ave.

Stout, Melancthon, 240 Wabash ave.

Stover, Frank G., 95 Dearborn ave.

Stowell, W. G., 880 W. Madison.

Swain, Edgar D., 65 Randolph.

Swain, Oliver D., 869 N. Clark.

Swasey, James A., 3017 Michigan ave.

T

Talbot, Eugene S., (M. D., Rush) Penn. Col. D. S., 1873; 125 State, room 23; 9 to 5. Prof. Dental Surgery, Wom. Med. Col.

Thompson, Jas. F., 69 State, room 16.
Towner, David M., 184 W. Monroe.

U

Upp, Chas. W., 279 State.

V

Vigneron, Eugene, 14 Central Music Hall.

W

• Wachter, C. H., 262 S. Halsted.
Wachter, John C., 279 Clark.
Wallace, Arthur C., 359 State, room 4.
Wassall, J. W., 208 Dearborn ave.
Way, James P., 336 W. 12th.
Webb, Morrell A., 858 W. Van Buren, 8 to 12, 1 to 5.
Wells, R. S., 713 Warren ave.
West, Geo. N., 70 Dearborn, room 7.
West, Wm. H., 243 State, room 32.
Wetterer, Albert, 40 Clark.
Wetterer, Herman, 40 Clark.
Whaley, John J., 216 N. Clark.
Wheeler, Thomas B., 257 31st.
Wikoff, B. D., 70 Dearborn, room 9.
Wilkie, Sylvester, M., 544 Blue Island ave., 8 to 5; tel. 9085.
Wilson, Harry H., Chi. Col. Dent. Surg., 1887; 692 W. Lake. Clinical Instructor, Central Free Disp.
Wilson, John J., 692 W. Lake.
Woolley, J. H., 69 Dearborn, room 16.

Y

Yates, F. G., 130 Dearborn, room 83.
Young, John H., 164 Dearborn, room 615.

Z

Zeno, Wm., cor. Lincoln and Cleveland, 8 to 12, 2 to 4; res. 1229 N. Halsted. Superintendent N. Chicago Dental Infirmary.
Zinn, Frank H., 496 W. Madison; tel. 7022.

Oakwood Springs Sanitarium.

OAKWOOD SPRINGS PARK FROM THE NORTH.

THE grounds consist of a beautifully wooded park of sixty-three acres overlooking Lake Geneva and the city, and commanding the most delightful and extensive views. It is, in its highest point, more than fifty feet above the lake, and presents a charmingly undulating surface, covered, in its entire area, by great oaks and magnificent forest trees. Its slopes, while nowhere abrupt and everywhere sufficiently gentle to permit of easy ascent, yet lead to valleys of sufficient depth to give the most delightful variety and pleasing effect to the landscape. The buildings consist of two principal hotels (the plans of a third have just been completed by the architect). The hotels are widely separated from each other, each having independent and extensive grounds. These buildings are not "hotels" in the common acceptation of the word, as they are not open to the traveling public, nor to the summer resorters. Apartments can be taken only by patients actually under treatment. Neither are they like hotels in construction, but outwardly have the character of grand private mansions, while in internal construction they are arranged so as to give to guests all the quiet and privacy of their own homes. This Sanitarium is exclusively for treatment of diseases of the Brain and Nervous System.

Please see pages 16, 121, 137, 170 and 183, for further information address

Oscar A. King, M.D., Supt., 70 Monroe St., Chicago.

COOK COUNTY DENTISTS OUTSIDE CHICAGO.

BARRINGTON.

Otis, Chas. B.

CICERO.

Pruyn, Chas. P., Oak Park; also Chicago.
Ritter, John M., Oak Park.

EVANSTON.

Bacon, D. C., S. Evanston
Bacon, M. T., S. Evanston.
Freeman, Clarence B.
Freeman, Henry A.
Garnsey, Chas. A. P.
Henderson, Luther D.
Timerman, Edwin C.
Whitefield, Geo. E.

HYDE PARK.

Barr, J. A., Grand Crossing.
Buell, Harry C., 3906 Cottage Grove ave.
Brown, James A., Pullman.
Cooper, John M., 408 Bowen ave.
De Pew, H. H., South Chicago.
Dayan, L. F., 118 53d.
Dryer, Levi T., Pullman.
Eshbaugh, Frank, 1370 Oakwood boul.
Eastman, G. E., South Chicago.
Freeman, Daniel B., 4000 Drexel boul.
Huskins, Jno. E., 126 53d.
Kester, Edmund W., 5101 State.
Lambert, Edmund, 3904 Indiana ave.
Lewis, Chas. W.
Marshall, Clara W., 4258 Cottage Grove ave.
Reedy, Geo., 5101 State.
Ullery, Frank B., 3949 Cottage Grove ave.; 3946 Drexel boul.
Withrow, H., 3904 Indiana ave.

LAKE.

Blackburn, Robert, 4251 S. Halsted.
Carson, Henry F., 503 63d.
De la Matter, Walter.
Fowler, J. W., 6204 Wentworth ave.
Kollock, Jennie C.
O'Brien, Henry, cor. Halsted and 42d.
Richards, W. P., 6218 Wentworth ave.
Witt, Wm., 4134 S. Halsted.

LAKE VIEW.

Becker, Geo. H., 1102 N. Halsted.
Bowman, Fred H., 1412 N. Clark.
Custer, Wm., 1202 Diversey ave.
Hermann, Richard, 1102 N. Halsted.
McIntosh, Edward M., Ravenswood.
Magnusson, Howard C., 41 Baxter, also Chicago.
Roberts, B. J., cor. Wrightwood and Lincoln aves.
Zeno, Wm., 1229 N. Halsted.

LYONS.

Carey, Warren, La Grange.

PALATINE.

Coltrin, Chas. W.
Olcott, Edwin W.
Persons, E. W.
Putnam, Rush M.

PROVISO.

Varney, Luther H., Maywood.
Wilson, Thomas L., Riverside.

COLWELL'S

LYING-IN INSTITUTION

3012, 3014 and 3016 Calumet Avenue.

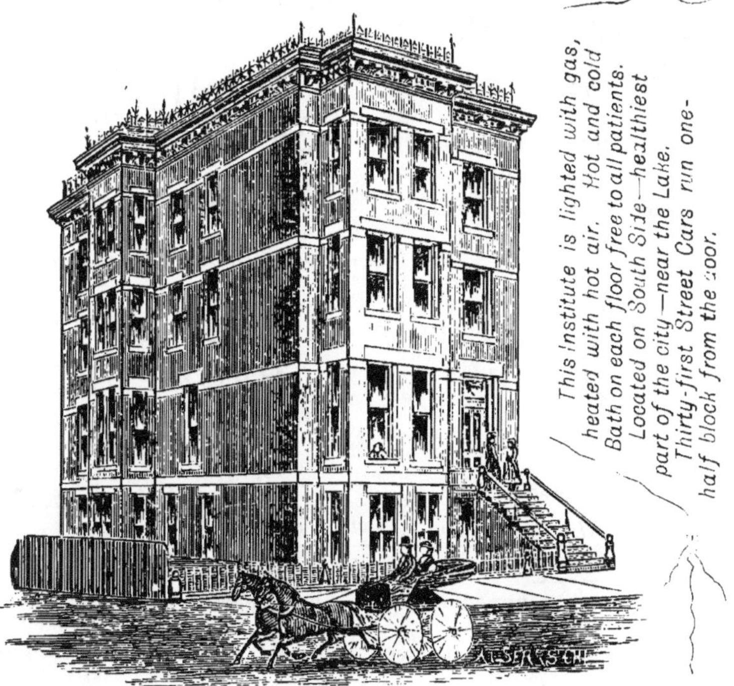

This Institute is lighted with gas, heated with hot air. Hot and cold Bath on each floor free to all patients. Located on South Side—healthiest part of the city—near the Lake. Thirty-first Street Cars run one-half block from the door.

SOUTH AND EAST VIEW OF BUILDING.

To Physicians

This Institute is for the benefit and PROTECTION OF UNFORTUNATE WOMEN during pregnancy and confinement; with every convenience for their comfort. Medical Attention and Trained Nurses at reasonable terms. Board from $10 to $25 per week. The price is governed by the room and length of time they remain. A home provided for the nfant if the parents desire to part with it. Physicians knowing respectable families wishing to take a child as their own call or correspond with me at office and residence.

B. L. COLWELL, M. D.,

TELEPHONE NO. 8432. 3414 CALUMET AVE., CHICAGO

PHYSICIANS ARE CORDIALLY INVITED TO CALL.

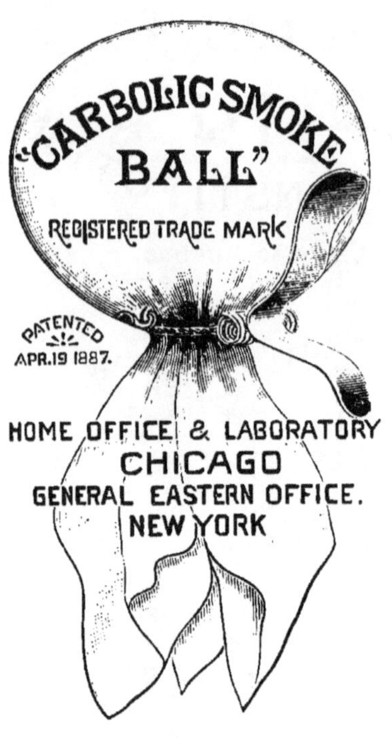

A GREAT SELLER!

Because it fulfills all promises.

THE

DRUGGISTS' BONANZA,

Because of its popularity, immense sales and clean cut profit.

"Carbolic Smoke BALLS."

$15.00 PER DOZ. RETAIL PRICE. $2.00.

"DEBELLATOR PACKAGE."

$6.00 PER DOZ RETAIL PRICE, $1.00

Where both remedies are properly used in cases of *Catarrh, Asthma, Bronchitis, Hay Fever, Neuralgia, Croup, Throat and Lung Affections* (consumption alone excepted), we warrant immediate benefit and a speedy cure.

We invite all reputable physicians to call at our office parlors and investigate our claims.

Druggists need not hesitate in warranting satisfactory results in every case.

It is a simple and pleasant treatment, speedy and sure, permanent and lasting results.

Beware of Imitations. Deal with Reliable Druggists.

CARBOLIC SMOKE BALL CO.,

S. W. cor. Clark and Madison Sts., CHICAGO, ILL.

Entrance 134 Madison Street.

CHICAGO DRUGGISTS.

(Retail).

Ackerman, George H., 851 W. Harrison.
Ackerman, Geo. H. & Co., 409 S. Western ave.
Adams, L. L., 225 31st.
Ade, Samuel G., 528 W. Indiana.
Ahlborn, Frank H., 395 Wells.
Allen, Charles B., 61 N. State.
Amundson, O. E. & Co., 804 W. North av.
Angear, W. J. S., 182 N. Halsted and 482 W. Lake.
Arend, A., 189 Madison.
Armbrecht, L., Jr., 724 W. Lake.
Armbrecht, Wm. H., 338 W. Madison.
Artz, O. W., 163 Webster.
Ashbury John, 962 W. 12th.

B

Bach, August, 251 Clybourn pl.
Baker, William, 2888 Archer ave.
Barber, Geo. L., 524 W. Madison.
Bardonski, Victor, 615 Noble.
Bartlett, N. Gray, 22d cor. Indiana ave.
Baseler, C. F., 445 North ave.
Basse, William H. T., 660 S. Halsted.
Bassett, Geo. R., 1324 Ogden ave.
Bate, Henry J., 44 Wells.
Baur, H. F., 180 W. Madison.
Beall, Geo. H., 750 W. Harrison and 310 Hermitage ave.
Becker, H. V., 292 N. State.
Beckwith, Henry J., 29 31st.
Behrens, Paul J., 727 W. Indiana.
Behrens, Theo. R., 505 S. Canal.
Berger, Frank J. 1486 Milwaukee ave.
Betting & Ross, 249 S. Western ave.
Bishop, Arthur S., 639 W. Madison.
Blahnik, Lorenz, 88 W. 18th.
Blood, I. W., 3100 Archer ave.
Bodemann & Conrad, 239 State.
Boeckler, C., 158 Willow.
Borcherdt, J. C., 735 W. Madison.
Borges, Leo, 189 Maxwell.
Borland, M. W., 378 W. Van Buren.
Boyd, S. J., 675 W. Lake.
Boysen, L. C., 210 Clark.
Brabrook, J. F., 160 W. Harrison and 666 W. Van Buren.
Brabrook, Walter A., 296 Ogden ave.
Braun, John A., 3100 Wentworth ave.
Brendecke, A. C., 56 W. Randolph.
Brendecke, J., 468 W. Chicago ave.
Breves, Charles, 275 W. Van Buren.
Brorson, N. C., 285 Noble.
Brunner, W. H., 1128 N. Harrison.
Bruuns, N., 286 W. Indiana.
Buchman, H. W., 2461 Indiana ave.
Buchman & Hirsch, 251 35th.
Buck & Rayner, State cor. Madison and 117 Clark.
Burlingham, L., 351 N. Clark.
Bush, F. C., 3659 S. Halsted.
Butler, F. A., 39th and Vincennes ave.

Button, Chas. E., 744 W. Van Buren.

C

Camp, A. R., 167 4th ave.
Campbell, L. J. M., 2763 Archer ave.
Caron, E. L., 109 Blue Island ave.
Casey, James, 130 31st.
Charbonnel, A. B., 3037 Cottage Grove ave.
Cheney, Lucian P., 237 N. Clark.
Christian, J. T., 1454 W. Lake.
Clancy, C. L., 525 W. Van Buren.
Clark Bros., 511 State.
Class, C. F., 887 N. Halsted.
Clinch, J. H. M., 325 Washington boul.
Coleman, John F. & Co., 208 N. Clark.
Colman, Martin A., 329 W. Van Buren.
Coltzau, O., 2169 Archer ave.
Conradi, Herman, 441 State.
Cowan, A. S., 100 N. State.
Cozine Bros., 3645 Cottage Grove ave.
Crain, J. W., 785 W. Madison.
Cunradi, Charles, Archer and Hart aves.
Cunradi, Julius, 2904 Archer ave.

D

Dahlberg, Alfred, 2459 Wentworth ave.
Dale & Sempill, Clark and Madison.
Davis, A. J., 169 W. Madison.
Dell & Co., 276 W. Madison.
Dessau, W. F., 430 W. Randolph.
Devendorf, C. A., Ogden ave. and Jackson.
Dietz, John, 2842 Butterfield.
Dietz, John, 3905 Wentworth ave.
Dinet & Delfosse, 286 Armitage.
Dodge, Fred W., 182 W. 12th and 482 S. Western ave.
Doherty & Marshall, 3458 Cottage Grove ave.
Doolittle Bros., 165 18th.
Doolittle & Co., 329 W. Van Buren.
Dougherty & Winter, 280 Clark.
Dow, G. Q. & Co., 118 53d.
Dreier, Emil, 259 Milwaukee ave.
Druehl, F. A., 422 W. 12th.
Druehl & Franken, 802 S. Halsted.
Duerselen, H. H., 635 N. Clark.
Dunham, W. F., 67 W. Van Buren.
Dunkel, Gustav, 363 W. Chicago ave.
Dyche, D. R., N. W. cor. State and Randolph.

E

Eakins, D. D., 362 Wabash ave.
Eberlein, Fred, 50 W. Madison.
Ebert, Albert E., 426 State.
Eldred, W. H. & Co., 47 Kinzie.
Ellefsen, B. J. H., 128 Oak.
Ellsworth, L., 351 Lincoln ave.
Elsner, Charles F., 1061 Milwaukee ave.
Emrick, G. M. & Co., 145 S. Canal and 467 W. Chicago ave.
Erich, Victor, 2500 Cottage Grove ave.

F

Falkenberg, F. E., 688 W. Chicago ave.
Fechter, A. E., 62 Canalport ave.
Feldkamp & Hallberg, 445 N. Clark.
Finckh, Wm., 212 Clybourn ave.
Fischer, E. J., 557 Sedgwick.
Fitch, Henry A., 884 W. Madison.
Fleischer, A. T., 296 N. Market.
Forsyth & Schmid, 3100 State, 31st and Wallace, and 162 22d.

Fox, Francis M., 3169 Archer ave.
Franck, August, 361 Blue Island ave.
Fredigké, C. C., 3500 State.
Freund & Williams, 258 31st.
Fry, I. H., 1800 Wabash ave.

G

Gale & Blocki, 44 Monroe, 111 Randolph and 126 N. Clark.
Garrison, A. L., 485 Clark.
Gazzolo, Frank, 490 W. Madison.
Goetz, Albert, 952 Milwaukee ave.
Goetz. Henry, 114 Clark.
Goll & Henkel, 661 W. 12th.
Goodman, F. M., 654 W. Van Buren.
Grassly, C. M., 287 W. 12th.
Grenamier, J. T., 439 W. Madison.
Guertler, Joseph, 439 W. Taylor.
Guy, G. O., 428 W. Van Buren.

H

Haanshuus, S. K., 241 Milwaukee ave.
Haeger, Fredk., 1595 Milwaukee ave.
Haller, Otto G. & Co., 1235 Milwaukee ave.
Halsey, Bros., 27 Washington.
Handsmann, F. P., 564 S. Halsted.
Hanke, Rudolph, 80 Chicago ave.
Harrison, H. B., 262 W. Randolph.
Harrington, J. J., 37 Rush.
Hartman, F., 350 Dearborn.
Hartwig, Chas. F., 476 Milwaukee ave.
Hasse, F., 570 Milwaukee ave.
Hattermann, C. F. 994 Milwaukee ave.
Hautau, Henry, 156 W. Randolph.
Heland, John, 113 W. Madison.
Heine, Geo. B. 3101 S. Halsted.
Heinemann. M., 211 Webster ave.
Heinemann, Paul L. I., 339 S. Jefferson.
Heller, E. H., 3022 Archer ave.
Helmich, A., 134 S. Halsted.
Hermann, E. Von, 420 26th.
Herron, S. P., 452 Wells.
Hess, John C., 247 Division.
Hesselroth, Lawrence, 107 Chicago ave.
Hewitt, T. A., 277 W. Monroe.
Heydenreich, Max, 231 W. Randolph.
Heylmann, C. J. 296 W. Lake.
Hibben, H. K., Michael Reese Hospital.
Higgins, J. D., 342 W. Lake.
Himming, Wm. A., 681 Larrabee.
Hinckley, H. A. 3209 Cottage Grove ave.
Hogey, J. H., 3038 Cottage Grove ave.
Holbrook & Warhanik, 726 W. Division.
Holroyd, E. A., 1014 and 1082 W. Lake.
Hooper, J. H., 171 N. Clark.
Holthoefer, H. J., 3160 State.
Hottinger, Otto, 869 N. Clark.
Hubbard & Gore, 3901 Cottage Grove ave.
Hurst, N. N., 3906 State.
Hutter, Robt., 620 Larrabee.

J

Jackson, S. H, 858 W. Van Buren.
Jacobi, A. M. & Co., 441 State.
Jacobson, Aug., 477 Ogden ave.
Jacobus, Judson, Indiana ave., cor. 31st.
Jamieson, T. N., 3900 Cottage Grove ave.
Janson, William, 426 N. Ashland ave,
Jarmuth, Henry, 87 29th.
Jauncy, William, 312 W. Indiana.

Jentzsch, R., 1086 W. 12th.
Josenhaus, R., 242 W. North ave.
Justi, W. F., Cal. and North ave.

K

Kadlec, L. W., 136 W. 12th.
Kadlec, L. W. & Co., 585 Center ave.
Kellner, M. G., 358 Larrabee.
Kennedy, C. C., 339 W. Harrison.
Kenning, R. H., 1333 W. Lake.
Kerr, J. B., 182 S. Desplaines and 537 W. Lake.
Kettering, Albert, 2603 S. Halsted.
Kiessling, J. C., 709 W. 21st.
King, E. H., cor. 22d and Wabash ave.
Kleene, F., 318 Milwaukee ave.
Klein & Sawyer, 322 W. Madison.
Klinkowstroem, E. V., Division and Hoyne ave.
Klotz, Aug., 426 N. Ashland ave.
Knight, A. P., 3300 State.
Knowles, Frank J., 475 Ogden ave.
Kraft, H. F. Co., 641 W. Madison.
Krembs & Co., 183 W. Randolph.
Kreyssler, C. E., 2614 Cottage Grove ave.
Krouskup, Walter H., 3659 Wentworth ave. and 3658 State.
Krzeminski, C. E. 150 North ave.

L

Lange, Ignatz, 849 W. Indiana.
Langenhan, Hy., 324 Blue Island ave.
Latto, Wm. W., 987 Ogden ave.
Lee, J. P., 260 S. Halsted and 527 W. Van Buren.
Leenheer, B., 871 W. 22nd.
Lehman, L., 2401 State.
Leusman, F. A., 2290 Archer ave.
Leyser, J. B., 381 Division.
Liese, F., 451 Larrabee.
Lion, John H., 176 and 178 W. Division.
List, F., 2724 State.
Litourneau, R. G. & A. E. 626 W. Harrison.
Livesy, R., 849 W. Indiana.
Long, Edward, 303 Blue Island ave.
Lundvall, August, 229 Division.
Lynch, John, 358 37th.
Lyneman, F. E., 660 W. Jackson.
Lyon, E. A., State, cor. 22nd.

M

McChesney, A. C., 660 W. Jackson.
McCormick, E., 771 Clybourn ave.
McDowell, V. P., 188 35th.
McPherson, E. K., 479 W. Van Buren.
McPherson, H. H., 530 W. Indiana.
MacDonald, B., 1049 W. Madison.
Macgee, Walter J. & Bro., 1035 and 1331 W. Van Buren.
Marbourg, J. G., 360 Ogden ave.
Mares, Frank M. & Bro, 2876 Archer ave.
Marr, W. L., 59 N. State.
Marshall & Dougherty, 3500 Cottage Grove.
Martin, G. P., 85 Lake.
Martin, Hugo W. C., 358 State.
Masquelet, John, 422 Clark.
Masquelet, S. F., 3454 S. Halsted.
Mathison, Soren & Co., 22nd, cor. Michigan ave.
Matthaei, Ernst, 13 Blue Island ave.
Matthei, Louis, 134 Canalport ave.

Matthews, Charles E. & Bro., 175 S. Western ave. and 247 Madison.
Maye, Julius, 675 W. Lake.
Maynard, H. S., 626 W. Lake.
Medcalfe, H. G. & Co., 72 N. Clark.
Mehl, William, 3601 S. Halsted.
Merz Bros., 189 Maxwell, 353 and 514 W 12th.
Merz, T. F., 657 Sedgwick.
Merten, Henry M., 172 Blue Island ave.
Metcalfe, H. G. & Co., 72 N. Clark.
Meyer, Albert, 2727 State.
Meyer, Fritz, 116 N. Center ave.
Miller, Gus. A. 799 W. Van Buren.
Miller & Co., 206 Milwaukee ave.
Misenheimer, E. C., 978 Harrison.
Mitchell, W. S., 722 W. Lake.
Moench & Rienhold, 146 N. Clark.
Monroe, F. LeB., 3501 Cottage Grove ave.
Moore, G. C., 519 Wabash ave.
Moore, W. B., 581 State.
Morrell, E. C., 690 W. Madison.
Morris, Wm. G., 835 W. Lake.
Muehlhan, Louis, 691 W. North ave.
Mueller, H. L. C., 307 Clybourn ave.
Mueller, H. W., 117 N. Wells.
Mullan, Eugene A., 722 W. 21st.
Musselwhite, A. C., 50 Clark and 92 Van Buren.
Myers, John, 1002 W. Madison.
McChesney, A. C., 660 W. Jackson.

N

Nehls, R. C., 3552 S. Halsted.
Neubert, Frederick, 2459 Wentworth ave.
Nichols, J D., 403 Racine ave.
Nielsen, Theo., 322 W. Chicago ave.
Nies, E. O., 201 W. Randolph.
Noll, Chas. F., 134 S. Halsted.
Novak, J., 723 W. 18th.

O

Oberman, A. N., 40 N. Clark.
Ohlendorf, H. L., 315 Wells.
O'Ryan, Ch. G. B., 248 Wells.
Otto, Edward, 113 Clybourn ave and 80 Chicago ave.
Otto, Emil, 649 W. 21st.
Overton, W. T., 564 W. Harrison.

P

Painter, D., 1412 Wabash ave.
Painter, E. J., 1355 Wabash.
Parke, George, 482 W. Lake.
Parsons, John, 194 W. 31st.
Patterson, T. H., 3640 Cottage Grove ave.
Peirce, Fred. D., 581 Ogden ave.
Pennington, D. & Co., 1056 Milwaukee ave.
Pierce, Frank W., Indiana ave., cor. 35th.
Pierron, J. J., 347 5th ave.
Plautz, C. H., 709 Milwaukee ave.
Pomaranc, M., 806 S. Ashland ave.
Poore, J. E., 3803 Cottage Grove ave.
Porter, M. N. & Co., 3858 State and 3900 Indiana ave.
Powell, Medford, 292 N. State.
Prince, James J., 745 S. Halsted.
Prince, Martin M., 573 Blue Island ave.
Propeck, J. W., 594 S. Canal.
Pyatt, Frank, 438 W. Madison.

R

Reichel, Augusta, 361 State.
Reinhard, Henry, 278 Wells.
Reuter & Co., 168 S. Halsted and 1369 W. Madison.
Rhode, R. E., 504 N. Clark.

Rice & Schaefer, 395 W. Harrison.
Richter, E. F., 136 Fullerton ave.
Ripke & Weber, 80 Webster ave. and 864 Clybourn ave.
Ritter, John, 372 W. Indiana.
Roberts, M. G., 201 W. Randolph.
Roby, Mrs. Ida, 177 31st.
Roemheld, J. & Co., 204 S. Halsted.
Rogers, Fred. C., 2412 Cottage Grove ave.
Rominger, Louis, 240 W. North ave.
Rosene, E. A., 318 Division.
Rosenfield, Samuel, 226 North ave.
Rosenthal, J., 440 W. Harrison.
Ross, B. L., Jr., 1 Lincoln ave.
Ross, J. W., 360 Ogden ave.
Ross, L. W., 1187 W. Harrison.
Rossmaessler, A. B., 224 Clybourn ave.
Rozienee, B., 2901 Wentworth ave.
Rudolphy, S., 389 S. Halsted.
Ruths, Geo. A., 947 W. Harrison.

S

Sargent, E. H. & Co., 125 State.
Schaar, J. G., 649 Blue Island ave.
Schaller, H., 224 Lincoln ave.
Schapper, F. C., 316 Sedgwick.
Schapper, Ferdinand, 757 N. Halsted.
Scherer, Andrew, 383 N. State.
Schick & Hess, 721 Elston ave.
Schimek, J. I., 547 Blue Island ave.
Schmeling, Ferd., 506 Wells.
Schmeling, Max, 388 Wells.
Schmidt, Fred. M., 7, 93 5th ave.
Schmidt & Fischer, 1558 Wabash ave. and 28 State.
Schmitt, H., 567 W. Chicago ave.
Schoen & Co., 147 W. Blackhawk.
Schrage, Frank, 977 N. Clark.
Schreiner, A., 311 North ave.
Schroeder, H., 453 and 833 Milwaukee ave.
Schroll, John M., 405 Larrabee and 412 Clybourn ave.
Scupham, W. C., 63 State.
Secord, F. G., 282 State.
Sethness & Co., 1218 Milwaukee ave.
Shannon, J. N., 34 S. Halsted.
Shauer, G. G., 631 Center ave.
Shean, John R., 378 Blue Island ave.
Sherman, Henry, 890 W. 21st.
Shugart, Jos., Jr., 947 W. Harrison.
Sieverman, B., 561 W. 12th.
Sill, Robert T., 628 W. Lake.
Slattery, Jas., 129 35th.
Smail, Jno. M., 654 W. Van Buren.
Smith, A. K., 206 W. Lake.
Smith, Junius J. & Co., 248 Clark.
Smith, Mrs. W. C., 349 Clark.
Smith, W. H., 192 N. Clark.
Somers, Frank G., 315 Chicago ave.
Sobey, R., 205 W. Adams.
Spork, Mrs. E., 322 W. Chicago ave.
Springer, C. B. & Co., 676 W. Indiana.
Squair, Francis, 567 W. Madison.
Stahl, E. L., 173 Van Buren.
Stamm, Andrew, 203 Blue Island ave.
Stamm, Carl, 2200 Archer ave.
Stangohr, Rudolph, 296 W. Division.
Starr, G. F., 571 S. Halsted and 2505 Archer ave.
Stevenson, Robt. & Co. 92 Lake
Stimming, William A., 681 Larrabee.
Stoddard, O. S., 771 W. Lake.
Straw, John I., 1007 W. Harrison.
Stringfield, C. P., 2448 Calumet

Studness, Otto, 455 Centre ave.
Sturges, C. S., 404 39th.
Sullivan, Thos. F., 3269 Cottage Grove ave.
Sumney, J. 281 Clark.
Swain, U. J., 262 W. Randolph.
Sweet, Henry, 126 and 234 Milwaukee ave.
Synon Brothers, 249 Blue Island ave.
Szwajkart, Adam, 699 Noble.

T

Tebbitts, F. M. & Co., 330 S. Halsted.
Templar, E. J., 250 Blue Island ave.
Tirrill, C. S., 100 S. Halsted.
Thayer, Fred A., 572 W. Madison.
Thiele, E., 1012 W. Lake and 2127 Archer ave.
Thoma, Henry F., 2011 Clark.
Thomas, J. B. 2700 Wentworth ave.
Thomas & Hess, 138 Wells.
Tobey, Richard, 205 W. Adams.
Trimen, J. W., 522 Wabash ave.
True, Charles J., 703 W. Harrison and 826 W. Madison.
Truman, A. S., 3640 State.
Truppel, Richard, 66 Wells.
Tucker, G. W., 1369 W. Madison.
Turnquist, A., 2424 Wentworth ave.
Tuthill, Robert, 128 Lake.

U

Uhlendorf, B., 2506 State.
Umsted, Geo., 384 W. Madison.

V

Van Buren, B., 1249 W. Madison.
Van Doozer, F. R., 129 35th.
Vanderpoel, C., 1160 S. Western ave.
Van Zandt, E., 34 S. Halsted.
Vaughn & Sawyer, 322 W. Madison.
Vaupell, Wm. R., 350 Loomis.
Venus, W. A., 781 W. 12th.
Vernon Bros., 323 S. Western ave.
Vogeler, A. G., 220 Lincoln ave.
Vogelsang, Robert, 254 Dayton.
Voltmer, Wm. J., Jr., 936 N. Halsted.
Von Hermann, Eugene, 2839 Cottage Grove ave.

W

Waiss, F, G., 372 S. Halsted.
Waldron, L. K., 189 Randolph and 448 N. Clark.
Walter, Jacob, Jr., 527 S. Halsted.
Warhonik, E. M., 736 W. Division.
Waxham, Charles, 3428 S. Halsted.
Weber, Ewald, 864 Clybourn ave.
Weinberger, C. M., 219 Wells.
Weir, R. M., 1235 W. Jackson.
Wells, James H., 951 W. Lake.
Wermuth, Wm. C., 117 Centre.
Werkmeister, M., 3459 State.
Wernicke, Oscar G., 109 Blue Island ave.
Wessman, A. J., 63 Chicago ave. and 81 Oak.
White Bros., 271 N. Clark.
White, J. W., 113 29th.
Whitfield, Thomas & Co., 240 Wabash ave.
Weidel, Ernest, 518 W. Chicago ave.
Wiese, Adolph G., 729 S. Halsted.
Wikoff, L. H., 342 W. Lake.
Wilson, Benj. D. F., 495 W. Madison.
Wilson, C. B., 783 W. Madison and 894 W. Lake.
Wilson, Julius H., 125 22d.
Wilson, W. R., 1257 Wabash ave.
Winholt, B., 208 W. Indiana.
Winholt, George, 2842 State.

Winholt, Theodore, 427 and 627 W. Indiana.
Winter & Baker, 280 Clark.
Wisshack, Geo. F., 186 W. Madison.
Wissing, Fritz, 752 W. 22d.
Wolfner, E. R., 351 Clark.
Woltersdorf, E. H., 900 W. 21st.
Woltersdorf, L., 171 Blue Island ave.
Woltmaer, Wm. J., Jr., 938 N. Halsted.
Woods, A. A., 943 W. Madison.
Wrixon, T. W., 515 Lincoln ave.

X

Xelowski, J. H., 709 Milwaukee ave.

Z

Zahn, Emil A., 1801 State.
Zender, M., 920 W. Van Buren.
Zindt, Julius N., 120½ Clybourn ave.

COOK COUNTY DRUGGISTS OUTSIDE CHICAGO.

BARRINGTON.

Abbott, H. T.
Waller, Louis.

CALUMET.

Schaffer, Ferdinand, Blue Island.
Schmitt, Emil, Blue Island.
Lowenthal, Louis, Washington Heights.

CICERO.

Lovett, La Motte, Oak Park.
Trail, R. H., Austin.

EVANSTON.

Bradley, W., 401 Davis.
Burbank, C. H., cor. Chicago and Greenleaf aves.
Garwood, M. C., 438 Davis.
Powell, M., cor. Chicago and Lincoln aves.
Rice, J. F., cor. Chicago and Greenleaf aves.
Schreiber Bros., cor. Lincoln and Chicago ave.
Williams, N. P., 414 Davis.

HYDE PARK.

Arnold & Merrill, 92d, cor. Erie ave.
Bickhaus, Conrad, Roseland.
Blood, Isaac W., 4108 Cottage Grove ave.
Bodemann & Conrad, Lake ave., cor. 50th.
Bonheim, Frank, 4847 State.
DuChateau, Arthur O., Torrence ave. N.E. cor. 106th.

Hermann, Jacob, 9200 Commercial ave.
Holcomb, T. A., E. Kensington.
Hubbard & Gore, 3901 Cottage Grove ave.
Jamieson, Thomas N., 3900 Cottage ave.
Jones, H. D. & Co., 132, 53d.
Lackner, Emil O., 4101 State.
Lawton, J., Grand Crossing,
LeBrett & Field, Rosalie, Music Hall block.
Macy, E. B. & Co., Ewing ave. cor. 99th.
McCann, J. B., Oakwood boul. cor. Cottage Grove ave.
Porter, M. N. & Co., 3900 Indiana ave.
Rehm, Julius, Grand Crossing.
Robinson, H. C., Grand Crossing.
Sandmeyer, A. L., 63d cor. Madison ave.
Secord, F. G., Pullman.
Schmidt, Ernest A., Colehour.
Sparks, Addison, Kensington.
Shaw & Pearson, Winnipeg block.
Sturges, Chas. S., 404 39th.
Terhune, B. F., 5059 State.
Wardner, Morton S., 4258 Cottage Grove ave.
Wheeler, F. A., Hegewisch.

LAKE.

Barnes, John, 825 43d.
Bell, John I., 4700 State.
Bonheim, Lee M., 5103 Wentworth ave.
Brooke, Firman C., 5108 S. Halsted.
Burgwedel, A. C., 6124 Wentworth ave.
Caldwell, C. P., 711 43d.
Curry, C. P., 6034 Halsted.
Deitz, John, 3901 Wentworth ave.
Hoeger, S. M., 4222 Ashland ave.
Hogen & Hisgen, 6218 Wentworth ave.
Holmes, H. H., 63d cor. Wallace.
Houghton, H. J., 6560 Wentworth ave.
Hurst, N. B., 3906 State, and 5100 Wentworth ave.
Kotzenberg, Charles, 4203 Halsted.
Masquelet, John, 4822 Ashland ave.
McDougal, William G., 4631 Wentworth ave.
Mehl, William, 4700 Ashland ave.
Minaker, W., 6204 Wentworth ave., Englewood.
North, Charles F., 63d and Wentworth ave.
Oliver, William H., Vincennes ave. near 81st, Auburn.
Perkins, J. S., 6801 Yale.
Porter, M. N. & Co., 3858 State, 3900 Indiana.
Reasner & Co., 5727 Wentworth ave.
Ritter, A. P., 4341 S. Halsted.

Rogers, Fred D., 6110 State.
Sandmeyer, H., 736 43d.
Schaefer, E. R., 4356 State.
Schmidt, Herman, 4132 Wentworth ave.
Stowits, E. T., 503 63d.
Terhune, B. F., 5059 State.
Werneburg, William, 654 68th.

LAKE VIEW.

Abbott, W. C., Ravenswood Park.
Baker, Geo. R., cor. Halstead and Noble ave.
Beinssen, Wm., 1410 Clark.
Bishoff, E. G., 420 Lincoln ave.
Brown, R. L., 886 Lincoln ave.
Copp, B. D., 722 Lincoln ave.
Ellsworth, Lewis, 351 Lincoln ave.
Fetherston, E. B., Ravenswood.
Frohn, John B., 876 Clybourn ave.
Jacob, Charles W., 1732 Ashland ave.
Klenze, William T., 1301 Belmont ave.
Luning, August G., Lincoln ave. cor. Diversey.
Nichols, Edward H., 403 Racine ave.
Pursell. Peter H., E. Ravenswood Park,
Richter, Ernst F., 136 Fullerton ave.
Schneider, T. B., 918 Lincoln ave.
Scholer, E. C., 886 Lincoln ave.
Wells, C. S., Clark cor. Diversey.
Wrixon, Thomas W., 511 Lincoln ave.

LYONS.

Higgins, Arthur, LaGrange.
Hillmer, W. B., LaGrange.

MAINE.

Carrier, C. W., DesPlaines.

PROVISO.

Clendenen, Irving, Maywood.
Moore, J. P., Maywood.
Van Tuyl, E. A., Riverside.

www.ingramcontent.com/pod-product-compliance
Lightning Source LLC
Chambersburg PA
CBHW030824270326
41928CB00007B/878

* 9 7 8 3 7 4 4 7 8 3 2 6 2 *